Arctic Genius

Arctic Genius

Sir William Edward Parry: The Original Arctic Explorer

Trevor Ware

First published in Great Britain in 2025 by
Pen & Sword History
An imprint of Pen & Sword Books Limited
Yorkshire – Philadelphia

Copyright © Trevor Ware 2025

ISBN 978 1 39903 821 8

The right of Trevor Ware to be identified as
Author of this Work has been asserted by him in accordance
with the Copyright, Designs and Patents Act 1988.

A CIP catalogue record for this book is
available from the British Library.

All rights reserved. No part of this book may be reproduced, transmitted, downloaded, decompiled or reverse engineered in any form or by any means, electronic or mechanical including photocopying, recording or by any information storage and retrieval system, without permission from the Publisher in writing. No part of this book may be used or reproduced in any manner for the purpose of training artificial intelligence technologies or systems.

Typeset by Mac Style
Printed in the United States of America by
Integrated Books International

The Publisher's authorised representative in the EU for product safety is Authorised Rep Compliance Ltd., Ground Floor, 71 Lower Baggot Street, Dublin D02 P593, Ireland.
www.arccompliance.com

For a complete list of Pen & Sword titles please contact

PEN & SWORD BOOKS LIMITED
47 Church Street, Barnsley, South Yorkshire, S70 2AS, England
E-mail: enquiries@pen-and-sword.co.uk
Website: www.pen-and-sword.co.uk
or
PEN AND SWORD BOOKS
1950 Lawrence Road, Havertown, PA 19083, USA
E-mail: uspen-and-sword@casematepublishers.com
Website: www.penandswordbooks.com

This book is for:

Agnes, Fergus, Hector, Jack, Josh, Maisie, Millie, Rafferty, Tom, Xan, Zac and Ziggy.

Contents

Acknowledgements ix
Introduction x

Part I

Chapter 1   Bath and Childhood   3
Chapter 2   Blockade   9
Chapter 3   An Object Peculiarly British   22
Chapter 4   A Pleasure Cruise   32
Chapter 5   Through Lancaster Sound   44
Chapter 6   'No Common Men'   51
Chapter 7   Fame and Family   63
Chapter 8   Illigliuk and Igloolik   71
Chapter 9   Fury Beach   82
Chapter 10   Marriage and the Stanleys   93
Chapter 11   Dark and Dismal Solitude   104

Part II

Chapter 12   A Thankless Office   115
Chapter 13   The Commissioner   123
Chapter 14   The Great Dividing Range   133

Chapter 15	Unfinished Business	142
Chapter 16	Steam Machinery and Packet Boats	150
Chapter 17	Departures	157
Chapter 18	Tragic Hero of Polar Exploration	166
Chapter 19	The Two Hospitals	170
Chapter 20	Epitaphs	178

Maps	185
Appendix 1: Geomagnetism	191
Appendix 2: Polar Geology	194
Appendix 3: Arctic Botany	196
Appendix 4: Meteorology	198
Notes	201
Bibliography	218
Index	225

Acknowledgements

Many different people have helped with the researching and writing of my book. Thanks to Anne and Johnny Roberts I was able to make contact with the descendants of William Edward Parry. Their help has been invaluable not only by allowing access to their family archive but also from showing me his telescope and sextant. There was nothing to equal the excitement of holding his most important navigational instruments, both in very good condition!

Naomi Boneham and Lucy Martin at the Scott Polar Research Institute, Cambridge have been typically patient and helpful making me copies of letters to and from Isabella and Sir W.E. Parry and their family members during the many weeks the archive reading room was shut due to Covid restrictions. I have been given valuable advice over the various maps of the polar region by Eugene Rae at the Royal Geographical Society Library, and by Emma Floyd at the Yale Centre for British Art about George Lyon's extraordinary watercolour sketches and private journal. Becky Sheldon at the Derbyshire Records Office has swiftly provided copies of the correspondence between Franklin and Parry from the Gell family collection. Henry Nolte helped me interpret the names and various coloured drawings of the plants collected on Melville Island and the Melville Peninsula, now in the Pollok Morris Collection at the Royal Scottish Museum. Beatrice Okoro at the National Maritime Museum was always ready to assist me with my tentative online searches and invariably discovered what I needed. Mike Maccoy gave me considerable IT help which I badly needed in preparing the MS, and my wife who has put up with my absences behind my computer and at various libraries and frequent references to Parry and his career!

And last, but certainly not least, I wish to thank Colin Whimster and his fine yacht *Sulasgeir*, without whom I would never have sailed up to a glacier and been swiftly inspired to investigate the many complexities of ice, the polar regions and the intermingled careers of Parry and Franklin, two of Britain's most famous and heroic explorers.

Introduction

During a sailing holiday with friends on the north coast of Norway we went into Holands Fjord to visit the Svartisen glacier, the second largest in Norway. It was an unusual and unique experience and imparted awe and apprehension from some distance away.

Our initial sensation was the rapid fall in temperature and then as it came nearer we saw its immense height rising 3,000 metres above the Fjord with splashes of vivid green and blue light glinting deep within crevasses as it plunged finally into the sea. As we left, the dense mist hanging overhead suddenly dropped like a curtain in a theatre, and the glacier vanished from sight. This was nature in an awe inspiring and unusually powerful state which left me absorbed so deeply that I became obsessed with the desire to know everything I could about glaciers and ice in general. How had this glacier grown so huge? How old might it be? What forces caused it to move? Why did it have such pure and vibrant colours below the surface? These and many other questions compelled me to read every book on ice I could and to find drawings and photographs of ice in picture galleries, online sources and travel accounts and journals. Eventually I decided to develop my interest much further by taking a research degree at Sussex University in 2013. The thesis I prepared was entitled, 'The Meaning of Ice. Scientific Scrutiny and the visual record obtained from the British Polar Expeditions between 1772 and 1854.'

Over the following five years I went on studying scientific reports, engravings from many travel books and descriptions of ice written during the Arctic and Antarctic voyages from the fourteenth century to modern times. There were the journals and sketch books made by Grand Tour travellers and the pioneer climbers in the Swiss Alps and I discovered numbers of engravings made during the 'Little Ice Age' in Britain which lasted from the mid-seventeenth to the mid-nineteenth century. These showed the very small amount of recorded knowledge that existed about the substance itself, the different forms it took and why, strangely, it never became salt laden in the sea. Most descriptions during the little ice age focused on the misery and hardship it inflicted on the poor and their animals but, very occasionally, there arose a strange euphoria among the wider population. One example being the ice fairs held on the River Thames and various rivers and canals elsewhere in Britain.

The emotional responses to ice are quite similar in the accounts of most eighteenth and nineteenth century travellers and authors, being succinctly summed up by the American writer Marjorie Hope Nicholson with the opinion that, 'ice-covered regions are worlds that puzzle, amaze, astound and enthral by their very differences to our own world'. These places, like the Norwegian glacier, provided imagined landscapes onto which painters, poets and authors of many nationalities could project visions of demons, giants, monsters, apparitions, folkloric heroes and heroines as well as pagan deities.

Modern scientific study of ice and glaciers started in the French Alps in the late eighteenth century, followed by intense exploration in the Arctic by the British Admiralty supported by the Royal Society of London, and in due course, the Royal Geographical Society. These bodies comprised many of the same individuals who agreed among themselves to fund fresh expeditions into the polar regions. One of the few remaining unexplored places in the early nineteenth century. These continued through to the early twentieth century concluding in Antarctica, first circumnavigated by Captain James Cook approximately one hundred and fifty years previously. The expeditions were equipped with every kind of scientific instrument to record in the smallest detail: wildlife, types of the ice, meteorological systems and the earth's magnetism and shape. They returned with great quantities of data including new surveys and maps, specimens of many different kinds and particularly evocative drawings and sketches of the ice, topography and wildlife.

Understanding ice found at sea was crucial to a long-standing but unaccomplished British national endeavour. Since Elizabethan times, when Martin Frobisher went looking for gold deposits and discovered Baffin Island in 1557, the British hoped to find a sea route between the Atlantic and Pacific to profit from the valuable spice markets of the Far East. If found this route would be much less hazardous and far shorter than sailing round the Cape of Good Hope or Cape Horn. The passage would greatly increase British maritime and scientific prestige internationally and thwart attempts by other nations, especially the expanding Russian empire with several settlements already in Alaska, from building more new military settlements in northern Canada. The Colonial Office and the Admiralty were equally determined to halt any such expansion which could quickly threaten the security of their colony. The unexplored Arctic was also of personal interest to several illustrious and wealthy members of the Royal Society most especially its President Sir Joseph Banks (1743–1820).

In its conception and for similar geopolitical reasons, the Northwest Passage expeditions can be compared to the NASA moon shots and landings of the late 1960s and 1970s by the United States of America. The cost to the British exchequer, estimated over more than thirty years, were at today's values similar when related to the distance in miles travelled.

The early expeditions had few reliable records or evidence with which to anticipate the direction and strength of the sea ice, driven by tides and polar winds that could easily destroy small wooden sailing vessels. Ships of the early explorers, Hudson, Davis, Middleton and Baffin, having been damaged or destroyed by icebergs, crushed in pack ice and by snowstorms that created thick ice on decks and rigging so rapidly they capsized. Only the masters of commercial whaling ships, such as William Scoresby Jnr. (1789–1857) had sufficient knowledge and experience to identify and predict dangers under different conditions and take avoiding action.

Arctic navigation was complicated because of the strength of magnetism found in northern regions which caused large deviations to compass readings due to the proximity of the 'magnetic' Pole. These phenomena, although anticipated by ship masters, had not been measured for strength and direction in different longitudes and latitudes precisely so that corrections to compass readings had to be approximate.

To measure the source and strength of this geomagnetism, new instruments were frequently devised and older versions improved, of which some were unusable or unreliable in sub-zero temperatures. All needed to be tested by separation from the magnetic field created by the metal on a ship, calibrated, and used in secure observatories.

More difficulties arose from the mirages and distortions of low temperatures and temperature gradients above the ice pack. These changed apparent distances, shapes of landmarks, and, under certain conditions, duplicated the suns and stars seen in the sky. Navigational instruments, such as sextants and telescopes, were awkward to use because glass and mirrors glazed over in very low temperatures and metal eyepieces froze onto the skin.

Expeditions were restricted to five summer months from mid-May to mid-October as winter darkness in the Arctic lasted twenty-four hours during November, December, January and February, with the sun permanently below or just above the horizon. Daytime temperatures could reach -40° Fahrenheit, and conditions for distant travel seemed impossible. The main reason why comprehensive surveying and exploration had not been attempted in Arctic latitudes.

A myriad of large and small islands, deep channels, inlets with dead ends lay inside the Canadian Arctic 'polynia' or archipelago. On the north side lay the polar icecap and to the south hundreds of miles of bleak tundra between Labrador in the east and the Straits of Bering in Alaska far to the west. Communication with the outside world within this labyrinth was impossible and while whaling ships might appear in the more accessible places for fishing during summer months, any expedition becoming lost or trapped had little chance of being found or rescued.

Voyages before 1818 had produced basic charts and maps with places discovered such as landmarks, and a few with depths and compass bearings, but most of them were inconsistent with one another and were not always composed using one method and annotated fully – James Cook's Australasian surveys and his charts for Newfoundland being notable exceptions. New expeditions needed to survey each mile of coastline taking regular soundings for depths, bearings on the most prominent topographical features, and providing clear sketches of their recorded position and outlines. This was arduous and difficult work normally undertaken in boats small enough to get close in to the shore. Propelled only by oars or sails they had to cover large areas around the base ship, while other overland expeditions had light boats or canoes sufficient for a few men with all their food, tents and other belongings, to journey down turbulent rivers, portage round waterfalls, and paddle miles across ice-strewn inlets and suffer terrible attacks by insects.

The British expeditions from 1818 were hastened forward by information, sometimes fictional, that the Russian Navy was exploring the same region from the west. By the late 1860s all this Canadian archipelago, the western coast of Hudson's Bay and the north-eastern North American peninsula had been mapped, given names and documented for most geological, botanical, ethnographical and mineralogical discoveries. The Arctic ice and its movements and concentrations though continued to be a mystery. Crews returned with accounts of its swiftly changing appearance, thicknesses and coverage all different from one year to the next. Only a century or more later, with better meteorology, surface radar, the electron microscope and global positioning systems (GPS), could some predictability be managed but not until after the RMS *Titanic* had struck a rogue iceberg and sunk.

The indigenous people living in the Arctic were willing to show Europeans how to use the ice, make clear windows, thick walls and even insulation for their houses or igloos. They knew its uses and characteristics but few British explorers took notice. Even today a misconception remains that the Inuit have four hundred different names for types of snow when they are in fact for ice!

The naval men leading the expeditions kept very short and concise ship's logs showing their exact position at hourly intervals. These were supported by special notes describing items of even minor scientific or geographic interest. The officers were untrained observational scientists and their journals and logs valuable guides for subsequent voyages, if sanctioned by the Admiralty. They were published for sale to the public while the reports concerning newly returned voyages were topical. These accounts fetched high prices and were usually oversubscribed. An inexhaustible interest continued for years for the engravings they used. The scientific appendices occasionally as lengthy as the journal itself, were less to the public taste. One officer whose journals with startling engravings and thorough appendices stands out from others were those published by Captain William Edward Parry. RN (1790–1855).

The mysterious disappearance of HMS *Erebus* and HMS *Terror* under the command of Sir John Franklin (1786–1846) is still remembered. Parliament authorised a large and imposing memorial in Waterloo Place just off Carlton Terrace in London, which commemorates each and every crew member by name. A recent discovery of one ship, the *Erebus*, well preserved on the seabed close to King William Island, has recreated international interest over the likely causes of the tragedy. One probable reason being the unexpected return of thick ice at both ends of the Peel Sound soon after Franklin decided to follow that passage to the south.

Parry, like Franklin, possessed strong stamina, energy and an extraordinary measure of personal bravery. He commanded four successive expeditions, pioneering equipment and the techniques needed for survival during winter above or near to the Arctic Circle. During all these only a handful of men died (some of whom had started with underlying health problems) but now has been forgotten except by his descendants, a few cartographers, maritime historians and polar scientists, despite during his lifetime being recognised throughout Europe as the 'father of British Arctic exploration'.

During his ten years in the Arctic he discovered a route into the Northwest Passage through Baffins Bay, mapped the northern and eastern coasts of Hudson's Bay and the Melville Peninsula of Canada and got closer to the North Pole than anyone had before or was to do so during the next half century.

His life was linked with that of John Franklin professionally and personally, they were close friends. Both had been young midshipmen during the Napoleonic Wars and held strong Christian beliefs as well as enlightened ideas over the welfare and education of sailors. The brutal conditions that prevailed for sailors on the lower decks of warships were anathema to them both. Parry, especially, worked hard to find ways to improve their education, health, moral welfare and financial security. He firmly believed that all sailors should be properly treated and that the Royal Navy would become a stronger force if this general attitude was taken. The men chosen for his expeditions signed up many times despite knowing the dangers they would face. Their common understanding was that he would be cautious about risking their lives and would always treat them as individuals. He fought many battles on the sailor's behalf against the neglectful, parsimonious and uncaring attitude of the senior officers on the Board of Admiralty.

After his Arctic expeditions Parry became the first Commissioner of the Australian Agricultural Commission sent out to New South Wales to manage their huge farming and mining undertakings. He arrived as smaller independent farms newly settled in the colony, as well as the finances of the AAC, were close to insolvency. After five years' resolute work he introduced improvements that had a lasting benefit to the settlers' quality of life and a sustainable agricultural and mining business for the London directors of the AAC. His ideas included the growing of suitable varieties of grapes for wine production.

Throughout the years abroad and for a shorter period in Norfolk, he was allowed to remain on the Navy list on half pay. The Admiralty decided they urgently needed his administrative skills and his positive approach to problems which could be put to immediate use reorganising and revitalising the loss-making Royal Mail Steam Packet services which had been transferred by Act of Parliament. At the same time he was commissioned to establish a new Admiralty department becoming Comptroller of Steam Machinery and responsible for the evaluation of new engineering machinery and applications suitable for warships. The leading designers, who were often the owners of engineering companies, went to Parry to seek his opinion and approval before any drawings or models were shown to the conservative and consistently sceptical members of the Board who preferred to live on with what they understood, the sailing navy and all its famous victories.

His final years were spent as Governor of Haslar Hospital near Portsmouth and then the Royal Naval Hospital at Greenwich. During those years he brought in overdue reforms and important changes warmly welcomed by patients as well as the medical staff. He went on lobbying continually for radical changes to help sailors conditions, including nutritionally improved daily rations, binding service contracts, pensions and basic safety rules.

He could not, though, find any way to stop the evil practice of flogging!

During his years at Haslar Hospital he and Lady Jane Franklin worked together to persuade the Admiralty to persist with the search for her missing husband. As the most experienced and successful Arctic officer alive he offered advice on where and how they should look for the expedition especially as for him there was a terrible and poignant truth. He had done his best to support Franklin's bid to lead the final expedition despite his age and relative inexperience of the conditions he would find. He had also sided with John Barrow and opposed the route which Franklin himself preferred and had been involved with fitting steam engines on both ships which he must have known were too underpowered to push through heavy ice.

Shortly before his death in 1855 he was promoted to Rear Admiral, the high rank that his friend Franklin would have reached a few months before him had he survived.

His life divides naturally into two parts and so does this biography. In the first part, during the Arctic years, he became a popular national and international celebrity, even though he had failed to discover the elusive Northwest Passage. In the second part, he endured an extraordinarily varied and mostly successful career and remained famous but increasingly committed to the conditions of the AAC settlers as well as the transported convicts and their families in New South Wales, and then again once more in England the sailors who had accompanied him on his voyages and others like them.

Members of the Parry family have written accounts of his life, extolling his strong Christianity, deep social conscience and consistent piety. I have tried, with the help of the family archives and letters written by contemporaries, to give a different perspective while not neglecting these admirable qualities.

He was a doughty explorer and a man who always wished to 'see for himself', however difficult, exhausting or dangerous.

His father had encouraged his interest in geology and agriculture and he grew into a remarkably good observational scientist. His upbringing gave him a taste for music, conversation, parties, and amateur theatricals. He was a loving and devoted husband to two wives, losing his first wife Isabella Stanley as a result of childbirth. There were eight children from that marriage, of whom four died in childhood, and two more with his second wife Catherine Hankinson. Both women were remarkable in their own right. He suffered badly towards the end of his life from rheumatism and stomach pains and was very short of money, the Admiralty having appointed him to several posts at once without paying the salary commensurate with the work. The conversions of his salary and emoluments are given throughout this account at 1917 values as recorded by the National Archives at Kew.

In March 1857, an article appeared in *Blackwood's Edinburgh Magazine*, a respected scientific journal, which described him as an 'Arctic Genius'. This accolade is no less true today but does not, I believe, adequately sum up all his many achievements. He was a modest and unassuming individual even after years of national and international celebrity which most certainly accounts for his being less well remembered after his death in 1855. The Royal Navy took years to accept reforms to which he had devoted both time and energy over many years seeking for all sailors and their dependents. For this, as well as the four pioneering Arctic expeditions he led, he should be far better remembered and celebrated.

Part I

Chapter 1

Bath and Childhood

William Edward Parry was born on 19 December 1790, the fourth son and second youngest of nine children born to Caleb and Sarah Parry. (Fig. 1.). The family lived in a small terraced house in Catherine Place in the centre of the city close to the Weston Road. He grew up like many children belonging to large families, with a natural gregariousness and lasting interest in all his family, attributes which remained all his life. One sibling, a boy called George Rodney, died before William Edward was born, whilst the remainder comprised his two older brothers, Charles and Frederick, eleven and seven years his senior, and five older sisters, Maria, Matilda (Matty), Gertrude (Gatty), Emma and Caroline who was two years younger than himself. Despite a wide discrepancy in ages the Parry children were all close to one another, due to the sociable, caring and devoted nature of Sarah.

Caleb Parry, his father, was the son of Joshua Parry, an ordained Minister in the Presbyterian Church who married Sarah Rigby, the daughter of a wealthy wool merchant. Caleb was raised in Cirencester, educated at Warrington Academy and went to Edinburgh University to study medicine, where he graduated in 1778. Before leaving the university he was elected President of the Edinburgh Medical Society at the very young age of 23. He then married Sarah, six years his senior, and after a short tour of France, Switzerland and Germany, they returned to England and settled in Bath.

Caleb was a gifted physician with a passion for medical research and a wide range of agricultural and scientific subjects. Sarah was sweet natured, charming, beautiful and unworldly. Characteristics which it was said she never lost right up to the day she died.[1] Her favourite child was her youngest boy William Edward, who steadfastly refused to be dressed in petticoats as boys at this time were sometimes made to do. He was large for his age, extremely boisterous with a good ear for music, and excelled at sport of all kinds. He was a pupil at the Bath Grammar School under the tuition of the headmaster, Dr Morgan; he was popular and academically above average, especially in mathematics. Sir George Markham describes him as 'a tall athlete, popular, with a good ear for music and a talent for acting'.[2] Fellow alumni at the same school were Sir Sidney Smith, Thomas de Quincey and Dr Lysons, the antiquary.

4   Arctic Genius

When Caleb and Sarah moved into the city it had been fashionable and famous throughout Europe since Roman times because of the therapeutic warm spring waters, the baths and more recently two famous Hospitals. Visitors, many seeking medical advice, nearly equalled the resident population which, at the census of 1803, numbered 33,000. Competition among the various medical practitioners was intense and Caleb needed to work hard to establish his name; he did so by travelling outside the City to find new patients and by joining both the Bath and West of England Society and the Bath Philosophical Society attended by many leading men of letters and science such as William Herschel the astronomer and Joseph Priestley the physicist. The two important societies offered him connections with wealthy, well-educated and influential people as well as providing a useful platform for giving medical papers on his discoveries and treatments. He was unusually amenable to collaborating with other doctors treating patients suffering from little understood medical complaints. His willingness to make examinations and give second opinions helped over time to develop his knowledge of unusual symptoms and conditions, their identification and possible treatments. One discovery, for which he is especially remembered, was the identification of angina pectoris, a hitherto untreatable heart condition with life-threatening implications on circulation of the blood and the lungs. His research on this was pioneering and he worked alongside his old school friend Edward Jenner, with whom he co-founded the Fleece Medical Society, named after the public house in which they met.

Dr Parry was elected to the Royal Society of London in 1800 with a modest and short citation, 'Caleb Hillier Parry MD, Physician at Bath a gentleman well versed in various branches of natural knowledge'. One of the fifteen Fellows who signed his election paper was Sir Benjamin Hobhouse FRS his brother-in-law.[3]

Caleb's determination and deductive skills as well as the remedies he proposed were sought by a large patient list who paid well for advice. This list also helped him identify and diagnose treatments for several unusual and rare conditions, including megacolon, polycythaemia and exophthalmic thyrotoxicosis.[4] His meticulous record keeping and willingness to give the poorer townspeople free treatment were replicated by the socially conscious actions of Ted, his youngest son.

An interest in sheep farming on the 60 acre farm on the northern side of the city where Caleb carried out experimental breeding to improve the quality of English woollens, a source of valuable exports and income for big landowners, also played a large part in William Edward's later life. Caleb wrote an essay on Merino sheep which won a prize from the Board of Agriculture in 1807. The *Bath Herald* of 21 October 1807 reported that George III had sent two

Merino Ryeland crosses to the Bath and West Agricultural Society for trials but Dr Parry had produced wool superior to the originals.

Another of his lifelong passions was for geology, then a new but increasingly popular study, becoming a founder member of the Geological Society also in 1807. Rock specimens and fossils of every kind were put in his pockets wherever he went and his sons were frequently exhorted to send him any discoveries from their travels and became William Edward's habit throughout life. Consequently, the Parry houses and not just Caleb's own rooms became stacked with samples. Jenner was also an avid collector and like Caleb, a balloonist!

As his reputation and wealth grew the family were moved to a far bigger house at No. 27, the Circus, the most fashionable part of the city and an address for medical men that compares to London's Harley Street. Using a legacy from Sarah's mother, he designed and built a country house on the northern side of the city near to his farm called Summerhill (Fig. 2.) From 1793 this large house was repeatedly enlarged but only lived in during the summer months to escape the heat, noise and tremendous bustle of the city.

William Edward, always called Ted by his family, grew up in a city full of famous and remarkable individuals whose discoveries and inventions included a range of different and new sciences. Men including Richard Beddoes, Humphry Davy the chemist, the astronomers William and Margaret Herschel, Edward Jenner Snr and Joseph Priestley. Caleb was on equal terms with these individuals not only because of his medical background but for an intellect that could interpret and interact with the new scientific ideas and opinions of his contemporaries.[5]

The 'sciences' as they were to become known were intrinsic to Bath's reputation not only for the enlightenment philosophers specialising in the natural world but also amongst people with literary reputations. Writers came to the famous, but short lived, Philosophical Society meetings, of which William Herschel and Joseph Priestley were leading members. The city was a natural magnet for other classes of people occupied in less worthy pursuits including actors, gamblers, medical quacks and many half-pay army and naval officers looking for social introductions that would lead to new commissions.

Jane Austen made long visits to Bath in the 1790s, eventually moving there in 1801 after her father George Austen retired. She wrote unflattering descriptions in her novels and letters about the city after she left in 1806, 'With what happy feelings of escape'.[6] Her novel *Persuasion* gives a fairly contemptuous view of Bath society which, she wrote, showed widespread dissipation notwithstanding the restrictive manners and conventions of the age, as well as rampant snobbery, and 'a certain dullness of habit, if not of dress'.

'Dear Doctor Parry' is mentioned both in her diary and letters along with the scornful observation that in her opinion, Bath was, 'the last refuge of half pay officers and retired civil servants'. Caleb's patient list, however, was above most of these people. He was approached by senior naval officers, such as Admirals Howe, Rodney and Cornwallis, as well as the Duke of Clarence, the High Admiral of Great Britain the Prince Regent and future King. He was popular among many of the aristocracy and attended to exiled members of the French Court including the Duke of Orleans and his family. Partly through these and other influential connections he was nominated as Fellow of the Royal Society of London, an organisation of established scientific and philosophical significance. Two of his sons, Charles and William, also became Fellows, and, like their father became intimate with the leading men of the day, including Sir Joseph Banks (1743–1820), its long-serving President and guiding spirit[7] The Parry family, including Ted, would have been familiar with the names of these people and there was as well the wide circle of friends and relations from both Caleb's and Sarah's families. Sarah had a busy social life spending regular periods away from Bath visiting relatives, usually accompanied by her eldest daughters. Charles accompanied their father on visits to meet Jenner and see patients living outside the city. Frederick, his middle brother, was busy with writing and poetry and Ted was left at home with Caroline, his youngest sister, and attended Bath Grammar school daily. He learnt to make up his own amusements, including playing the violin and acting. During school holidays he was sent to stay with his Aunt Amelia, married to Sir Benjamin Hobhouse who was a banker, barrister and President of the West of England Society.

A short description of his elder sisters is provided in the letter sent by Dr Burney to his daughter Fanny Burney, the novelist:[8]

July 22nd 1799. In bed at Dr Herschel's, half past five, where I can neither sleep nor idle. My dear Fanny, here I arrived at 7 o'clock. When we had conversed about ten minutes in came two other sweet girls about the same age (from 15 to 17), the daughters of Dr Parry of Bath on a visit here. More natural, obliging charming girls I have seldom seen; and moreover very pretty, I found they were musical and in other respects very well educated.

Charles and Frederick were away from home during much of Ted's childhood, attending a private school in Greenwich run by Dr Burney, then going up to Cambridge University. They travelled over to Gottingen University where they joined walking holidays with the Wordsworths, William and Dorothy, and Samuel Taylor Coleridge. Charles completed his medical studies and went back to Cheltenham in 1809 and married. It was hoped he would follow Caleb into his Bath practice but Charles disliked the city and rejected the idea. He helped Edward Jenner instead to whom he became almost like a son and as a qualified doctor and pathologist, he supported the case for vaccination against smallpox. After his time at Gottingen Frederick left Cambridge and, lacking any positive direction for his future career, was sent by his father to manage his newly acquired investment in the Myconi Creek project in British Guiana, or Guyana as it is known today.

Ted matured quickly with unusual self-confidence and self-sufficiency. His father worked extremely long hours, and he was either alone or just with Caroline over quite long periods. She recalled him as:

> having great aptitude in acquiring and retaining knowledge. His love of music and excellent ear for time and tune were also manifest. I heard his mother say, that, at four years old he would catch any air hearing it only once and that he would sing 'Rule Britannia' with all the spirit and energy of a man.[9]

Ted became restless and bored during the several summer holidays he stayed with his Aunt Amelia and Uncle Benjamin Hobhouse.[10] He had become a handsome boy with golden curls and a tanned complexion, tall, athletic, and grasped every opportunity for adventure or physical activity – including such sports as riding on his father's sheep when taken up to the farm.

Although performing well in school, especially in mathematics, it appears as though he was not to follow his elder brothers through university and into a profession. At some point he must have felt his future would be too circumscribed if lived in the 'dullness' of Bath Society as described by Jane Austen. As the object of his mother's and elder sister's adoration, with a constant round of parties, visits to relatives and the invitations from the important patients of his father, perhaps he craved escape and adventure.

These reasons or others led him to decide, aged 13½, that he would join the Royal Navy.[11] War with France had been in progress during much of his early childhood, during which the Royal Navy won famous victories under Admiral Nelson, including the Nile, Cape St Vincent and Copenhagen. The navy was highly regarded by the nation and when Admiral Cornwallis, who visited Bath to take the waters and visit Caleb, was introduced to him, it was his niece Elizabeth who suggested that Ted should become a sailor and enlist as a Volunteer First Class. This being a lowly position created to assess the suitability of new recruits before acceptance to the official rank of midshipman.

Sarah accepted her favourite child's decision, despite neither Caleb's nor her own family having any naval connections or history. Several circumstances may have helped her decide and acquiesce. Her husband's financial problems were growing due to his eagerness to speculate in property in and around Bath. There were repeated losses from the sheep farm as well as an unexpected failure of the Myaconi Creek ranch in British Guiana, into which he had invested capital from his mother-in-law's estate after she died. Running two large houses, the Circus and Summerhill, were contributing to their difficulties. The elder boys had been sent to expensive private schools and Cambridge but by 1803, Caleb may have been unable to afford any financial commitments of the same kind.

Eldest son Charles had been carefully educated to succeed him in his practice, while Frederick had been sent to Guiana. Ted may have been thought unsuitable for the calling of a doctor or an Anglican Minister, often the path chosen for a junior son. The family were church-going Anglicans and Ted was brought up a practising Christian but Caleb and Sarah both came from nonconformist backgrounds and several of their friends, such as the Fry family, were Quakers.

Neither parent at the time apparently believed the church offered a part in their youngest son's future. The church could never satisfy his enthusiasm for physical activity and thirst for adventure and experience in the wider world.[12] His uncle, Sir Benjamin Hobhouse, Secretary to the Board of Supplies in Henry Addington's Government and MP for Grampound in Cornwall, knew that his energetic young nephew possessed qualities valuable in a naval officer and gave his support to the prospect of his joining up. His opinions would have been influential with both parents.

Accompanied by one servant to help with his luggage, he travelled by coach to Plymouth on 17 January 1803, never having seen the sea and with scarcely a couple of hours on pleasure craft on the River Avon to inform him about life on the restless ocean.

Chapter 2

Blockade

The Channel Fleet was stationed outside Brest Harbour and along the Brittany coast to prevent the French home waters' fleet joining with the Spanish Navy and their Mediterranean squadron. The blockade began in 1795 during the French Revolutionary War, and was imposed by the British Government to enable them to avoid risky and expensive army operations in support of Royalist led rebellions on the west coast of France and was intended to give the rebels tactical support, rescue aristocrats fleeing the mob, and to stop French warships crossing the Atlantic to attack the sugar islands of the West Indies, a major source of British wealth. Henry Dundas, Secretary of State for War (1794) pronounced that, 'Complete success in the West Indies is essential to the interests and I will add, to the contentment of this country.'[1]

As the conflict continued these objectives changed and the blockade by the Channel Fleet became vital to Britain's survival. Forty French ships of the line stationed at Brest were the nucleus of a powerful invasion force. If unchecked these ships could intercept shipping going up the English Channel carrying much needed food and raw material supplies for Britain. The naval dockyard in Brest also used large quantities of timber, iron for armaments, cordage and food. Much of this needed to be moved by sea. The blockade could prevent these crucial materials reaching the port and so limit the seaworthiness and readiness of the French fleet.

Parry went to join the *Ville de Paris*, a large warship with 110 guns and flagship of the Channel Fleet commanded by Admiral Lord Cornwallis. Captain Ricketts was captain of this ship and was now his master as well as his tutor. It was a most fortunate start to his naval career since he was serving under two very senior officers whose attitude towards new recruits was enlightened compared with many of their contemporaries. Cornwallis did not share his predecessors' jaundiced views about midshipmen of whom many were young men of high birth, some with parliamentary connections or having senior Naval officers as patrons and relatives. His attitude being that 'young gentlemen', the term by which the midshipmen were generally known, were not to be accused indiscriminately of being part of a group labelled by both Jervis and Nelson as, 'representative of a vast overflow of young nobility which makes for rapid strides to the decay of Seamanship as well as Subordination'. Cornwallis' predecessor, John Jervis, Lord St Vincent (1800–1803), had reinforced the Channel Fleet and imposed a close inshore blockade in the spring of 1800. This was more hazardous than the 'offshore' or distant blockade due to the proximity of rocks and shoals, and attacks by smaller enemy ships with good local knowledge. This followed the collapse of the second British parliamentary coalition, the formation of the Armed Neutrality of the North and the invasion of the Electorate of Hanover by France which was clear evidence of her continuing aggressive ambitions. St Vincent was a veteran Admiral and a strict disciplinarian. His captains were reprehended for the smallest mistakes, any deviation from the given course or position, and even minor lapses of seamanship. Each ship had to tack and wear in close order at least once every night. The captain and first officer had to be on deck. Jervis got up every morning at 3.00am to see which of his captains was on deck first.

The blockade was never relaxed whatever the weather, and his determination was so absolute as to ignore Admiralty 'advice' that he should withdraw to the shelter of the English coast during winter gales. This defiance was at risk of causing serious damage to rigging and sails, as well as 'diseases of debility', and worsening of morale and discipline among his sailors.

His solution to these was the great attention given to the frequency of fresh foodstuffs, water, post and war materials brought out from Plymouth to Ushant by a large fleet of small ships to resupply the large men-of-war as well as all the frigates, sloops and raiding vessels. Some of these smaller vessels patrolled close inshore for more than twenty weeks without a rest and the physical demands on their crews, as well as the seaworthiness of their ships, was immense.

A decision which improved the sustainability of the blockade was a daily issue of lemon juice added to the sailors' rum ration that had been found effective against scurvy, a debilitating condition affecting mariners in the seventeenth and eighteenth centuries. This antiscorbutic had only recently been issued to the Channel Fleet as previously lemon juice was dispensed only on foreign stations. Channel Fleet captains were worried at any sign of scurvy and St Vincent, who had commanded the Mediterranean Fleet insisted and got from the Admiralty the cost of a daily issue of juice for everyone under his command – the direct result of which substantially improved the health and morale of the sailors when Parry joined.

Jervis' instruction to all the officers was "rub out can't", and put in "try", and his constant search for improvement in the fleet as well as a high level of local autonomy granted by the Admiralty, were among the reasons for only six French men-of-war escaping from Brest during his sixteen months in charge. The officers he led were imbued with this determination, professionalism and an unequivocally positive attitude. An active engagement with each captain reviewing performance and ships readiness were all the sign of an effective leadership which Parry could see and were the example he followed in the future.

Jervis' health was being affected partly as a result of his age, he was 70, but mainly from the stress of command, sleeplessness and the discomfort of staying at sea in all weather. He was recalled to the Admiralty to become the First Lord of the Admiralty in February 1802.

The Honourable William Cornwallis inherited from Jervis 'an organisation and a standard to work to'. The new Admiral was a successful and respected officer with several victories over the French to his credit. He had a more unassuming manner and fewer of St Vincent's abrasive and autocratic methods. He was less intent on looking for faults in his officers and had a tolerance for the recreation of his men even though he expected the same high standards as his predecessor. Admiral Collingwood, who served under him, described him as a 'man of very reserved habits and manners', and, 'of very few signals and fewer words'. He allowed officers under his command to evangelise seamen, including his flagship the *Ville de Paris*, generally known in the navy as 'blue light' ships.[2]

Parry joined as a Volunteer First Class when the Channel Fleet was back in Plymouth for refitting and repairs. During a pause in the blockade resulting from the short lived Treaty of Amiens signed in October 1801. On 18 May 1803 this treaty collapsed and the fleet went out once more to Ushant to resume its position further out to sea but, 'always near enough for the purpose.'[3] It remained on patrol in all weathers for three more years!

Whilst he benefited from Caleb's professional standing and contacts with men like Cornwallis the social standing of his family was not high and he was without many of the social advantages which could help move him rapidly up the list for promotion.. His solution was to work hard at seamanship, astral and lunar navigation, astronomy, plane trigonometry, and the duties of the 'mates man', including setting up complicated sets of rigging and sails.[4] One anecdote told by Ricketts, his captain, was about his eye for detail and readiness to suggest improvements that would improve the speed and handling of his ship.

After a friendly disagreement with Ricketts about the rigging of the main yard, Parry rapidly constructed a small scale model to prove his theory and his captain accepted that he was right, acknowledging himself fairly beaten. This story reflected well on both Parry's ingenuity and Ricketts' integrity. Parry learnt quickly and was made a midshipman after only two months at sea. Naval regulations at this time proposed at least two years.

He worked hard at mathematics and draughtsmanship, both of which came naturally to him and were key parts of the examinations to become a lieutenant, but, apart from the theoretical studies he followed, there were many commands and watch disciplines he had to learn to instruct the sailors and keep their respect, despite his young age.

Throughout his three long, usually tedious years in the Channel fleet battered by the Atlantic with its ferocious winter storms, living in wet and woeful conditions below deck, short of natural light, always crowded and full of unpleasant human odours, he learnt patience and self-discipline. Even with constant noise from straining timbers, bellowed orders, flapping canvas and the howling and turbulent winds of the Atlantic, he still managed to study. Even in a ship the size of *Ville de Paris* the best accommodation it could provide for its young gentlemen was in the orlop, the lowest deck above the hold and one other even darker space forward of the mizzen mast, called the cockpit. In both these places light came from tallow dips whose oily smokiness joined the stench of bilge water, rotting food and wet clothes. Both were below the waterline but preferable to the cramped conditions among the ship's main company on the gun decks and messes. Meals, except on those special occasions when invited to dine with the captain, were monotonous and nearly inedible.

'Young gentlemen' were appointed officers-in-training and included all midshipmen, and masters mates who were set above ships boys and the seamen. They were expected to show leadership qualities and prove their suitability for the privileged positions they occupied. They were expected to follow the etiquette and traditions of the other officers on the quarter deck. Boys under 11 by naval regulation were not allowed to join the navy although the rule was often ignored. James Anthony Gardner began his career on HMS *Conqueror* at the age of 5. Some 'gentlemen' were much older so that the average age was usually between 16 and 19. Depending on the size of the ship ten or more midshipmen could be undergoing training at any given time; the *Ville de Paris* had thirty. After two years' study midshipmen were eligible for promotion up to a master's mate position while waiting the captain's decision to appoint them as 'acting' lieutenant, if not yet old enough to sit their examination. Only the Admiralty could appoint a full lieutenant so Parry's next promotion to lieutenant and then 'first' lieutenant were highly important to his career. At the higher rank he would be the senior watch keeping officer and second only to the captain.

As he was still young, Parry lived in the gun room, and it was where he took his daily lessons instructed by some officers and the chaplain. The master gunner had his wife on board who acted as a surrogate parent to him and some of the other young men. Food was the same as that for the lower deck and the warranted officers, 'at breakfast we get tea and sea cake; at dinner we have either salt beef, pork, or pudding.' More affluent young gentlemen kept their own provisions and sent letters home like boarding school children today, pleading for food hampers. No letters exist to show if Parry asked for such things but he did get regular invitations to dine at the captain's table or in the officers' wardroom.

In letters to his parents Parry mentions breakfasting with both Cornwallis and Ricketts, and dining with Lord Cornwallis again a few weeks later.[5] His chief pleasure on both occasions having been the serving of soft bread called 'soft jack' or 'soft tommy'.

Cornwallis encouraged friendly and positive relationships among his officers with all the midshipmen. He urged his own son during his naval career to 'always be on good terms with his superior officers', and Parry applied himself readily to this principle, understanding that any midshipman's authority among lower deck seamen came from their perception that they were officers and gentlemen and were respected by the senior officers. Seamen knew that an order given by a midshipman must be obeyed without question. Commissioned officers on large warships being comparatively few the captain relied upon the 'young gentlemen' to uphold authority among as many as 700 men. Discipline as well as trust in the senior ranks throughout the Royal Navy 'rested more on persuasion than force'.

The Admiralty had quite recently recognised the midshipmen's role and ordered that uniforms should be issued to all, no longer wishing they wore any clothing they wished, and instructed that dress swords were always to be worn when dining with the captain. Special orders had been given in 1756 and 1759 emphasising the midshipmen's need to 'know the men assigned to their gun crew and be responsible for their presentation and professional readiness'. Despite these orders most decisions lay with individual captains, both as to exact responsibilities and the levels of authority given. In a letter to his mother several years later Parry wrote:

> the entrance of a boy into the Naval Service is, to speak my mind, rather a melancholy thing – for I know how in so dangerous a school whether he turns out well or ill. Few parents know how great a lottery it is and very few like to be told so.[6]

His enthusiasm for this new life appears to have been complete. Writing to his parents he confessed, '*no* life can be so pleasant as at sea', and again a few weeks later, ' I am very happy', after first thanking them for their twenty-one letters which had arrived all together! This satisfaction is partially explained by his description of 'loading more bullocks from HMS *Plantagenet*' and in great expectation at the prospect of 'going to take a peek [in Brest] at the French Fleet today'. Both events a most welcome interruption from the monotonous days patrolling far out at sea. One of the many letters from his parents told him that his brother Frederick had died from hydrocephalus and his emotions are very strong as he valiantly tried to console them by stating his conviction that he, Frederick, will have found 'eternal' life.[7]

After this letter was sent the fleet went into Brixham harbour to collect water because, as Parry wrote, 'the French cannot come out in this wind'. He tells his family that the crew are not allowed ashore for fear of desertion and that 'the *San Josef* was so short of able seamen she could no longer remain on station'. The strain on the men was growing despite the measures taken by Jervis and Cornwallis and because of a shortage of replacements. The *Ile de Paris* needed to be refitted at Portsmouth in August 1804 and he was allowed a few days leave but the general expectation was that Napoleon would invade England the following year.

The build up of newly constructed French warships at Brest and Toulon continued, and a large Spanish fleet remained inside Cadiz harbour so that all the enemy ports had to be kept under constant watch. The combined French and Spanish fleets amounted to 102 ships of the line, while the British had seventy. By September 1804, Cornwallis with his forty-four ships, including supply ships, had seven line of battle ships at Rochefort, seven at Ferrol, and sixteen watching Brest where twenty-six French ships of the line were anchored. There were no British reserves and Cornwallis was ordered not to take unnecessary risks during winter gales but, like Admiral Jervis, he kept every ship out at sea, despite the risk of being blown onto the terrifying rocks of the Brittany coast. (Fig. 3. The Wreck of the *Magnificent* on the Boufoloc Rock. 24 March 1804.)

The French fleet at Toulon commanded by Admiral Villeneuve left the Mediterranean, collected some Spanish ships at Cadiz, and sailed across the Atlantic to the West Indies, hoping to weaken the blockade by drawing off its forces. Nelson followed them but the French strategy failed and the fleet inside Brest under Admiral Gantheaume – who had twenty-one ships of the line, outnumbering the British – failed to escape to join forces with Villeneuve.[8] Cornwallis' contribution to the ensuing victory at Cape Trafalgar on 21 October 1805 was significant but the officers in the Channel Fleet felt deep disappointment at missing the greatest sea battle of the Napoleonic War. Frank Austen, Jane Austen's brother and an officer in the Mediterranean fleet, wrote to her:

> to lose all share in the glory of a day which surpasses all which ever went before, is what I cannot think of with any degree of patience, but, as I cannot write upon that subject without complaining, I will drop it for the present, till time and reflection reconcile me a little more to what I know is now inevitable.[9]

John Franklin, then a junior signals officer on HMS *Bellepheron*, just escaped death from sharpshooter fire from the French ship *L'Aigle*.

Despite his excellent record and a promotion to Vice Admiral of Great Britain, Cornwallis was removed by the new First Lord of the Admiralty, Charles Grey.

Before leaving his command he wrote to Caleb about his son:

> in regard to any civility I may have shewn young Parry, I never knew of anyone so generally approved of. He will experience civility and kindness from all while he continues to conduct himself as he has done, which, I dare believe, will be as long as he lives ... He is a fine steady lad. It is almost a pity he had not been to sea sooner, for he will I am sure, be fit for promotion before his time of servitude is out.[10]

The victory at Cape Trafalgar allowed the blockade to be reduced and Parry was given his first real leave for three years over the Christmas of 1805. The family were all present, except his sister Caroline, who had been sent away to boarding school for being 'too wilful', and of course Frederick, who had died. It was, Parry conceded later, a 'melancholy affair'.

His eagerly awaited examination for lieutenant was affected by a time when promotions were postponed and the qualifying age for candidates lowered to 19. This created a large number of candidates with fewer vacancies, as well as a number of 'passed' midshipmen and mate's masters waiting their first commission. Parry now saw that transfer into a smaller ship, preferably a frigate, would help him advance his claim to higher rank even before taking his examination when this was permitted. He was still only 16 years old.

'Young gentlemen' from frigates were an elite; set apart from the others they obtained advantages in their initial postings and sometimes faster progress with their careers. Through influence by Caleb and a lucky encounter with a Captain Baker in Portsmouth, Parry was transferred to the thirty-six gun frigate HMS *Tribune*. Baker was an old friend of the Parry family and *Tribune* part of a squadron patrolling close inshore and carrying out nuisance raids on French harbours. In one raid Parry was appointed prize master, bringing to England a small captured merchant ship loaded with wine and sardines. This was the kind of action he craved, even though over the next two years *Tribune* was chiefly engaged on reconnaissance and message-carrying duties. Parry was kept busy with navigation and drilling the crew and admitted such in a letter to his parents: 'I find however that I have not on the whole so much time of my own as when I kept watch; for now I cannot be sure of a minute in which I am not liable to be sent for on a hundred occasions.' During this time hoping for further action against the French, he writes of his desire to do his duty as a

'Christian and an officer. I shall always carry in my mind who is my Protector and my friend.'

In May 1808 Captain Baker was promoted and commissioned to command HMS *Vanguard*, flagship of the Baltic fleet under Vice Admiral De Saumarez and Parry went with him. They were ordered onto convoy duties, which between 1806 and 1810 were essential to protect British merchant ships targeted by the French to weaken the British war effort. A total of 2,210 ships, mostly in convoys, went through the Great Belt in the six months between June and December 1809. The naval historian N.A.M. Rodger in *Command of the Ocean. A naval History of Britain 1649–1815* notes that 'there were no losses'.

Despite increasing political pressure from the French, a few European nations continued trading with Britain, using their own merchantmen and others provided by the allies, all carrying supplies between British and overseas destinations. *Vanguard* was to protect essential materials such as timber, iron and pitch from neutral Sweden, sailing through the Kattegat across the North Sea and into British shipyards. The Danish Government, allies of the French and seeking retribution for their heavy defeat at Copenhagen, sent flotillas of shallow draught gun boats to attack the British convoys. Sometimes they managed to board and capture merchant ships before they reached the North Sea and as a way of countering this threat armed gunboats were despatched from *Vanguard* and other escorts. Parry was involved in these encounters.

On 3 January 1810 he passed his lieutenant's examination and two weeks later received his first commission, aged almost 20. He was posted to another frigate, HMS *Alexandria*, under Captain Quilliam, an officer who taught him seamanship skills later to prove especially useful to his Arctic career. He was given his own cabin, which had enough space for him to practice his violin three or four hours a day, and on which he was becoming accomplished. He was six feet two inches tall with a slight stoop caused by his years on ships with low headroom, had wavy chestnut hair cut short and brushed over his ears, and unusual grey eyes. Although admitting to being a passionate man he claimed to be well in control of his emotions which his letters generally confirm. He was thankful at leaving Captain Baker's ship because of that officer's violent and unpredictable temper. He also expressed contempt for decisions taken by the Admiralty which he considered led to inefficiency and loss of effectiveness. Some he shared with his close family but mostly kept them to himself.

Alexandria was part of the Baltic fleet but not regularly on convoy duty and was being used to protect the Spitzbergen whale fishery, then a valuable economic resource for Britain. Part of their orders were to prepare accurate coastal surveys with hydrographic readings, landmarks, marked hazards including reefs, shoals and rocks, and to update and improve all existing charts for the Shetland

Isles, Spitzbergen and the south-western coast of Norway and north-western Denmark. Parry had a natural aptitude for this work and, on his own initiative, also produced a pamphlet entitled, *Nautical Astronomy*, privately published by Caleb in 1816. Two years were spent on these duties, interrupted with a few convoy protection voyages against marauding Danish schooners and gun boats. Navigation close to the Norwegian coast was difficult and without reliable charts many shoals and hidden rocks concealed numerous enemy gunboats.

Pilotage close inshore was dangerous but it gave Parry considerable experience of manoeuvring in shallow and confined waters. In the spring of 1812 the *Alexandria* left Cromarty Firth in Scotland to go to 76 deg. north above the Arctic circle, where they were to map Bear Island between Spitzbergen and the North Cape of Norway. Before reaching Bear Island they were surrounded by floating ice, forcing them to steer an indirect and difficult course into strong winds and tides. Eventually they gave up and went eastwards to Norway. This was Parry's first experience with large formations of sea ice and seeing the sun above the horizon at midnight.

The French Wars were now in their eighteenth year. In recent years poor harvests in Britain and reduced imports created severe shortages in grain supplies needed for the Army. French trade embargoes, including sanctions collectively known as the 'Continental System', halted shipments to Britain and Spain from North America which were critical for Wellington's army. A sympathetic understanding between France and the new American Republic, as well as disputes over the 'pressing' of American sailors, led to renewed hostilities between the United States and Great Britain. A major proportion of the grain needed for Wellington's Peninsula Army came from the Unites States and the American Federalists, deciding that Napoleon would win the war, had cut off supplies where they could.

Sir John Warren commanded a small British squadron based at Halifax, Nova Scotia. He was ordered to protect British Canada, blockade the US Coast, and negotiate for large quantities of grain supplies from American merchants. Warren's orders were also to board all American ships trading out of New England, safeguard any neutral ships willing to move grain supplies to Spain, and attack and destroy American privateers.

In addition to these tasks he was to source his own rations at Halifax, a remote and not well supplied maritime outpost of Canada. The orders were so contradictory and the resources so few that it became farcical.

Lieutenant Parry sailed across the Atlantic for the first time and wrote:

the sight of the full moon just about sunset in these latitudes is one of the most sublime I ever saw: the clearness with which it is seen is, I suppose, to be attributed to the rarity of the atmosphere. With a brother officer, who, like myself knows only that he knows nothing, we have gone hand in hand together in our occupations. We have been going through Euclid again. He plays the flute and we have regular duets together: astronomy, mechanics, and chemistry have not been neglected.[11]

He joined HMS *La Hogue*, commanded by Captain Coote, and became part of another blockade at the eastern end of Long Island Sound, where they were ordered to raid up the Connecticut River using six small boats and to 'cut out' twenty-seven enemy privateering ships. His success was rewarded with a bar to his general service medal and in June 1814 he was posted to HMS *Maidstone* when the *Hogue* returned to England. He was 'second on the list for promotion to first lieutenant while still attached to the North American squadron'.[12] The news that Napoleon had escaped from Elba is rapidly considered as being likely to speed up the rate of promotion in the Navy, but not to everyone's benefit. There were 420 midshipmen and many lieutenants anxiously waiting for promotion![13]

Letters from Parry at Halifax where *Maidstone* was based, contained reports about the gypsum quarries, hydrous calcium sulphide, used to prepare Plaster of Paris and to improve agricultural soil fertility. Seven samples of this rock were despatched back to Bath. Halifax, he wrote, has 'benefited greatly from the war with America, and exports one hundred thousand tons of gypsum annually'. This material and other different specimens for father, and brother Charles now living near the family home in Bath, shows their continued passion for minerology and geology.

Their combined collections and appraisals of the various properties, sources and different uses helped develop the overall Parry reputation in local 'scientific' circles and with members of the Royal Society of London.

War with the United States ended at the Peace of Ghent in 1815, and Parry now had spare time in Halifax. He welcomed this despite the town having no 'musicality', and lamenting that there was no one to practice or perform with him on the violin. The port was 'ordinary and dull', but he made several friends. One being Charles Martyr the Admiral's secretary, whose close relationship Parry said, marked the advance and development of his religious principles and lasted for the next ten years when Martyr died prematurely. After one year he was made up to first lieutenant and given command of HMS *Carron*, a small twenty-gun sloop lying in Bermuda. He proudly described his ship as 'a very desirable command', even though the humidity and heat in Bermuda, 'usually between 82 and 84 degrees in the shade, creates terrible languor that is inconceivable', and admits that his books are, 'idle on the shelves of the captain's book case'.

Bermuda's population was growing rapidly. After 1815 and the ending of the British-American War of 1812–15, it became a free port. The consequence was more trade converging on St George's and Hamilton Harbour as American cargoes bound for the British West Indies had by law to be trans-shipped in Bermuda and sent onward in British ships. A new naval base and a large dockyard gave employment for many shipwrights, sailmakers and sailors and a local prize court sat when required to decide on awards from captured ships. The town of Hamilton offered most of the social pleasures Parry had been missing and his friends there included the Markham family with two cheerful, attractive and musical daughters, and with whom Parry clearly enjoyed many evenings for concerts and parlour games. These were a greater attraction to him than time spent with other officers among whom gambling, heavy drinking and card games were the usual pastimes. At 26 he was a serious young man with a strong Christian belief, a love of music, interested in topics of science, and poetry.[14] He mentions a romantic attachment to one Markham daughter but nothing resulted. He also stayed with the Boyard family and tuned their violins 'to their delight'.[15]

Parry's Christian belief and interest in music, science and poetry were discussed with George Robertson, a friend from England who made a welcome visit to Bermuda. 'One of the few naval officers to retain their classical education so as to make it subservient to useful purposes.' Robertson brought with him a copy of Caleb's recently published book on pathology and Parry immediately writes to his father praising the work and news that Caleb is intending to publish a new treatise on arteries.

In a letter home longer than usual he remarks that his 'intellectual stock is dismally poor', blaming distractions from his command, ill-discipline among the crew, and a need to be constantly on the '*qui vive*' from early until after dark. The crew of the *Carron* regularly petitioned him to be sent home and 'paid off', and his description of these men is highly critical. They are, he states, 'the refuse of other trades' and having an 'altered conduct from wartime'. He reports to Admiral Griffiths, his commanding officer, that he believes real seamen 'at the beginning of another war may never to sign on again' because of poor pay, poor conditions and bad treatment such as flogging for relatively minor offences. It is very evident that the degradation of British seamen began to affect him considerably from this time and he decided to try to improve their moral conduct, education and self-respect.

Hoping to attain quickly the next rank of commander, Parry considered applying to join the British Caribbean Squadron, where, because yellow fever was endemic and produced a high mortality among officers, promotions came more rapidly if you survived. He asked his father to place him on yet another 'list' for a different overseas station, but soon after was ordered by Admiral Griffiths to escort Lord Sherbrooke, the new Canadian Governor of General, to Quebec on HMS *Niger*. He was to be first lieutenant under Captain Jackson and as he was being rapidly transferred across to the *Niger* a parcel of books and new 'fashionable' clothes arrived from home, to which he managed to respond with a quick letter by return expressing, 'a thousand thanks for these and the list of the names of all my nephews and nieces'.[16] Long absence from home with one brief period of leave over Christmas 1808, meant that many of his nieces and nephews were unknown to him.

His visit to Quebec included two events, both described in one long and passionate letter to his parents. The 'romantic wildness of its [the St Lawrence River] shores', and their 'indescribable grandeur and sublimity'; and the other, a new steamboat sent down from Quebec to tow the *Niger* upriver because they had been becalmed. A 'truly wonderful piece of mechanism which goes at nine knots and is six hundred tons'.[17]

Parry decided to write to Bermuda where Admiral Griffiths was based hoping to persuade the Admiral to give him a more important command; unfortunately Griffiths had left for England by the time his letter arrived so he wrote to Caleb asking him to plead on his behalf and to deliver it in person when the Admiral came down to Bath. Soon afterwards he changed his mind and asked Caleb not to intervene as he had charge of Griffiths' son, a new midshipman on *Niger*, and believed this would be sufficient to help his interests, and that 'his reward would inevitably follow'. Nothing did result however and for the next nine months he continued to be discontented, restless and under-occupied in Halifax. A state he hated so much that he considered whether he should stay on in the Royal Navy or join the Merchant service instead.

His unhappiness was greatly magnified by exceptionally poor morale on the *Niger*. Lack of regular pay, a shortage of well-trained seamen, and frequent desertions chiefly 'by the useful men', caused him to complain bitterly in a letter dated 7 August 1816: 'I have to bully and coax the men, and plod on as well as I can.'[18] The faults, he considers, lie entirely with the Admiralty. As for the men he says:

> I cannot describe them as *seamen*, they are ragged, dirty, ignorant and given to thieving. They are given two weeks advance of wages, including a bed and blankets which they cannot do without. They get *no* pay abroad so the advance has to last for three years on a foreign station. Twenty-seven men have deserted the ship in Canada and it is necessary to turn her into a *prison*. The men who deserted from this ship were all useful seamen whose loss I have too much reason to regret almost every hour of my life.

From this moment onwards he became more determined to bring improvements to both the pay and conditions on the ships he commanded, and to force the bureaucracy at the Admiralty to recognise the benefits of such reforms.

Very near the top of the list for promotion, he hears from Sarah that Caleb had had a stroke and is semi-paralysed. He is given immediate compassionate leave and returned to England on the next packet boat, arriving home in Bath in November 1816.

Caleb's condition prevented him practising medicine or finishing and publishing the notes on arteries and other medical complaints assembled over many years. He could no longer manage the family farm or his complicated property affairs. Charles was able to help with most of their parents' problems meanwhile and since Ted had been away so long, he had very few suggestions or ideas to contribute. Combined with his uncertain prospects in the navy, this was a difficult and unhappy period in his life. Few of the meetings at the societies and scientific institutions in Bath, or the social circles of the family, retained much appeal. He was restless and anxious but one day reading a newspaper announcement about the naval expedition being fitted out to explore the northern regions he decided to write to an old family friend Sir Murray Maxwell, asking whether he would lobby the Admiralty on his behalf. Within a few months his future life and naval career changed dramatically for the better.

Chapter 3

An Object Peculiarly British

When Parry returned to England in November 1816, the French Wars had been over for two years but the state of the country was still far from normal.[1] Far-reaching social and economic changes had begun, including the rapid switch towards agricultural mechanisation which intensified a shift in population from countryside to big cities and towns, and provoked widespread demonstrations against the oppressive Corn Laws. There was also agitation amongst the urban population in sometimes violent protest over their desire for wider enfranchisement and overdue reform of the Parliamentary voting system, both of which had been in need of change for thirty years. Voters were disenfranchised unless they were property owners, and there remained a fair number of MPs representing very small or even non-existent boroughs. Adding to these problems were the number of men discharged from the armed forces now without either work or pensions, who resorted to petty crime and smuggling to make a living.

Bath, where Parry immediately went to see his paralysed father, had undergone many changes during the previous fourteen years. The city and its population had grown to 67,000 – twice the number it was when he left. Many new houses had been built in the suburbs as well as in its fashionable streets and crescents, bringing with them new shops, businesses, places of entertainment and different classes of professional people. His elder brother Charles, a doctor, was now living with his family at Summerhill while Sarah and Caleb had decided to move out of the Circus into a smaller house at 7, Sion Place. Parry was back in time to help them. In spite of his poor physical condition Caleb was still able to dictate articles and essays to Charles on the medical topics for which he was now famous. He was still visited by ex patients, including famous living admirals and members of the aristocracy.[2]

Parry did not spend long at home despite his extended absence and Caleb's declining health. His career was at a precarious stage and competition at his level of rank for worthwhile commissions leading to promotion was very intense. The opportunity to eventually command a ship of the line would much depend on this next step. 'To be an Admiral demanded early Captaincy, and there was little chance that a boy would make an Admiral at an active age unless he had reached post-captain in his twenties.'[3] Parry was now 26.

Continual pressure from Parliament over the size of the naval budget meant the Admiralty had been forced to make large reductions to its peacetime complement of officers and seamen. By 1817 its wartime strength of 140,000 seamen had reduced to 20,000. Ninety per cent of its officers were now on half pay with only a few who had retired or been pensioned off. This resulted in a ratio of one officer to every three seamen, which was unrealistic in peacetime. Promotions were very difficult to obtain for junior officers such as Parry without very powerful connections. Those that were available in peacetime scarcely offered the same opportunities for recognition and promotion as in wartime. Parry and Franklin, both on half pay, decided to apply for an expeditionary voyage using a steam-powered ship up the Congo River, despite knowing that tropical diseases in West Africa could end their lives abruptly. The Royal Navy in peacetime now had one increasingly valuable purpose which had been impossible to fully maintain during wartime.[4] This was the preparation of accurate hydrographical data containing detailed information about ocean depths, reefs, shoals, currents, tidal directions and magnetic compass deviations everywhere around the world.

The role became increasingly vital as Britain's expanding trade with its protective naval forces took ships to all the oceans of the globe. The recording and measurement of magnetic variations affecting ships' compasses was especially critical to these records. One more recent survey was for the coastline of southern and eastern Australia undertaken by Matthew Flinders in 1801–03. This survey had repeated the methods used by James Cook twenty five years previously and, despite serious risks to its crews, was of major importance to British colonisation in Australia. Franklin, a cousin of Flinders, was a member of this expedition which helped make his career. (see Fig. 8)

The Royal Society of London, who had sponsored Cook's pioneering voyages into the South Pacific, including New Zealand and Eastern Australia, now wished to continue new expeditions to increase their knowledge in specialist sciences and to discover territories of commercial value for the British Crown. These sciences such as botany, ornithology, minerology, geology and astronomy were emerging swiftly. Members of the Society and in particular Sir Joseph Banks its remarkable President, now agreed with the Board of Admiralty to part fund another search for the Northwest Passage and to try to locate the exact position of the North Pole.

The route between the Atlantic and the Pacific Oceans, either by crossing the region around the North Pole or following the North American coastline, had been probed by explorers for 250 years.[5] If a new shorter route was found the sea passage between Europe, the East Indies and China, could be reduced by months. This saving in time and distance would benefit the trading outlook with valuable new markets in the Far East, as well as avoiding the dangerous storms off the Cape of Good Hope and in the Straits of Magellan. This commercial advantage would be enough in itself, but long held rumours dating as far back as Elizabethan times maintained that gold deposits lay somewhere deep in the Arctic archipelago whilst other new factors had arisen which accelerated the authorities interest in starting exploration.

There was a new threat of Russian intervention along a sea route from the Pacific through the Bering Straits along the coast of Alaska and across northern Canada. One mission led by Lieutenant Kotzebue of the Russian Navy between 1815 to 1818 had discovered the Barents Sea blocked by thick ice most of the year as Cook had also discovered in 1776. Kotzebue made little progress along the Alaskan coast where Russia already occupied territory but the news alarmed the British Government who wanted no foreign presence in this empty region or for any discoveries rivalling those of Cook and Admiral Vancouver. The whole vast territory, apart from its few Russian settlements and the presumed but uncharted 'passage', was considered now to be a 'peculiarly' British endeavour.[6]

Sir Joseph Banks (1743–1820) the energetic and long-serving President of the Royal Society had accompanied Cook to the South Pacific in 1771–72. He had many personal scientific interests placing him at the hub of new scientific research and progress during an unusually active period of geographical discovery. Banks was a powerful figure with complete financial and personal independence and corresponded worldwide with many institutions and individuals on matters of science, manufacture, politics and trade. He was also wealthy and an efficient administrator and organiser; Banks was willing to initiate scientific 'experiments and interests' at home, as well as to encourage new expeditions overseas. Among his many attributes was an ability to persuade other wealthy individuals, including royalty, to support his pet projects and ideas. (See Fig. 6.)

Over several years Sir Joseph had been contacted by landowners in the north of England to ask for possible causes for the continual cold and wet summers affecting harvests. This had increased bread prices, reduced living standards for the poor, caused several public disturbances and was a great additional cost on the Parish Relief scheme. These conditions, many property owners feared, could lead to widespread civil unrest and potentially a revolution similar to that in France thirty years previously.

Several letters written by William Scoresby Jnr, an experienced whaling ship owner, had arrived reporting that during his whaling voyage in summer 1817, 'the ocean for 2,000 leagues, approximately 1,500 square miles, in the Greenland Sea between the parallels of 74 deg. and 80 deg. north, was perfectly clear of ice'. Scoresby suggested that this could be a reason for the failed harvests and harsh winters in Britain over the previous three years, because the heavy ice formations had moved far further south affecting the climate. He proposed an expedition to the West Greenland sea and up into the northern part of Baffin Bay to study the ice pack and whilst there to search for any route leading into a north-west passage.

Banks had a particular interest in the Arctic. As a young man he had joined a small ship exploring the north coast of Labrador and, like other adventurers, had been surprised and amazed by the size, colours and variety of icebergs sculpted into wonderful shapes, as well as the unusual mammals and birds, and the ethereal colours of the Aurora Borealis shimmering over the huge and empty landscape. Scoresby's science was entirely self-taught, but his extensive knowledge of the Arctic was evident from his drawings and observations, backed up by his accurate and meticulous records. He impressed Banks who realised that his experience of ice and the strange meteorological phenomena it produced would be for any expedition a great advantage. Scoresby sketched different types of ice crystals and snowflakes in detail describing the *parasalena* created by refracted light from over the icepack given certain conditions. Some of these, he showed, distorted distances and the shape of objects.

The interest Banks had for the Arctic was matched by another individual nowadays recognised as the true architect of British polar exploration during the first half of the nineteenth century. John Barrow, (1764–1847) the second Secretary to the British Board of Admiralty who visited Labrador and Greenland when only 17. (see Fig. 4)

Barrow had enormous influence on day-to-day operations of the Royal Navy, including appointments, naval policy and its working relationship with the Foreign and Colonial Office. He was a skilful manipulator working behind the scenes rather than as direct contender for authority and power. First appointed by Henry Dundas, the First Lord Melville, in 1807, Barrow had spent several years in China and Capetown as private secretary to Lord McCartney. He held strangely rigid notions about the Arctic and even composed verse about the frozen north similar in style to that of the poet James Thomson's 'Winter' from the famous work, *The Seasons*.[7] A hard working, political and energetic civil servant whose skills also included a shrewd appreciation of party politics. He became after many years of experience very capable at managing expectations and often conflicting opinions of the members of the Board of Admiralty. His only equal was the first Secretary to the Board who represented the naval interests in the House of Commons and had therefore to be an MP. The Board all had personal interests and chose individuals for commission from amongst men whom they had either fought alongside or joined to them by social or family connection.

The second Secretary explained new proposals and the reason for naval budgets, reported on the naval shipyards and ran the Navy Board. His experience, astuteness and perceptive political skill made Barrow indispensable to each First Sea Lord who usually changed with every new Government taking office.

On his retirement after forty-one years in office, the citation to Barrow signed by many officers whose careers he had promoted, including Parry, read: 'a man of talent, zeal and energy unceasingly displayed in the promotion of Arctic discovery'. His complete understanding of Admiralty appointments and the men who could be considered 'useful' officers with varying degrees of compliance and capabilities, has led one historian to name them all, 'Barrow's Boys'.

The confidence Barrow enjoyed with Banks continued up until the latter died. Both were empiricists in the widest sense, new scientists and influential members of the Royal Society, aware of, if not conversant in, all the rapidly expanding areas of new research. Barrow also became a founder member of the Geographical Society (1832) which gradually took on responsibility for most of the organisation and funding of land expeditions through north-western Canada.

Barrow's portrait shows a face with short cropped hair, bristling brows and a rather moon-like aspect. It also indicates his keenness of mind, energy and single-minded determination. One person who came to resent his great influence described him as one for whom:

> vanity and desire for distinction and importance will carry his patronage far beyond humble merit. He may serve a friend occasionally but his first care will always be for himself and the GREAT. He is not, however, to be neglected as the next best thing to making him a friend, is NOT making him an enemy. His bustling speculating spirit has got him into a great reputation in the Admiralty.

All this according to John Wilmington, the brother-in-law of John Franklin. In one letter to her husband Franklin's second wife Jane Griffin, having met John Barrow for the first time, wrote that by reputation he was, 'said to be humerous [sic] and obstinate, and that he exhibited both propensities.'[8]

Barrow used his position to promote both naval and diplomatic reasons for scientific research in the Arctic along with extensive new exploration and survey work. During his life he wrote over two hundred articles in the *Quarterly Review*, a Tory journal published by John Murray, as well as a similar number for the *Encyclopedia Britannica*. These sometimes show a vindictive side to his personality for he could and did undermine and destroy the reputations of the men who did not live up to his expectations, or disagreed openly with his ideas. One officer to receive this treatment was Captain John Ross, the officer commissioned as Parry's senior on his first Northwest Passage expedition in

1818. Barrow's family motto was '*Haec Neminisse Jurabit* – He will swear that he will never forget this.' In the case of John Ross, he never did.

A third individual important to Parry's career was Robert Saunders Dundas (1771–1851), the second Viscount Melville. He was made First Lord of the Admiralty on 25 March 1812, remaining in post until May 1827 – an unusually long tenure. With Barrow, who had worked for his late father and had been appointed by him into the Admiralty, the Arctic expeditions were safe. Henry Dundas, Barrow maintained, had, 'by his great exertions and prompt measures met the French threat, and defeated them'. His son, Robert Viscount Melville, has been described as a man who 'never made an enemy, nor lost a friend'. His authority and good working relationship with both Barrow and Sir Joseph Banks meant that the revitalised Arctic project planned for 1817 was endorsed by the Prince Regent, the future George IV, and the majority on the Board of Admiralty. Barrow and Melville shared the same opinion that patriotic and popular employment for the Royal Navy during peace time would be a good way of resisting Parliamentary efforts to reduce its budget. It would help retain some of the best officers and seamen, as well as showing how Britain continued to lead the world in scientific and maritime affairs.

Since Parry was staying with the Hobhouse family in London his uncle, Sir Benjamin Hobhouse, was able to arrange an invitation to meet Sir Joseph over breakfast. Postings for the West African expedition had been already been made and so responding to the advertisement for suitable men for an expedition to the north, he had written to Murray Maxwell, a friend with good naval connections. His letter said he was ready for 'hot or cold'. Maxwell had shown this letter, together with a copy of Parry's short pamphlet on astronomy to Barrow and both were sent over to Viscount Melville with a recommendation that Parry could be 'the man for such an appointment'.

Caleb's reputation as scientist and doctor clearly helped his son at this point. Sir Joseph told Parry on meeting, 'how glad he is to be acquainted with the son of Doctor Parry of Bath for whom he entertained the highest regard'.[9] After breakfast he showed Parry a map which he kept, showing the areas on the east coast of Greenland Scoresby had reported free from ice. This evidence he felt might account for the much reduced summer temperatures and higher than normal rainfall in Britain over the past three years.[10] Banks expressed his opinion that recently arrived cargoes of whale oil, when tested, were found to be very inferior in quality. The fall in sea temperatures from ice melt causing composition and combustibility of the oil to deteriorate and made for problems for most households for whom candles and whale oil were used for lighting. A twenty shilling bounty, about £50 today, per barrel was paid by the government on sperm oil as it was essential for lighting these homes.

The changes in climate were eventually found to be caused by the eruption of Mount Tambula in Java, which released millions of tonnes of thick ash into the atmosphere causing a warming effect over the northern hemisphere.

Banks offered to lend Parry some of his books and maps and provided him with an invitation to spend time in his private library whenever he wished. Parry's next letter to his parents was full of excitement about this meeting, telling them that Sir Joseph was 'committed to a real desire to do everything which can in the smallest degree, tend to the advancement of Science', and that they had a strong liking for one another.[11] Soon afterwards he was asked to meet with Barrow, who explained the plan for two different expeditions.

These two meetings were, he writes, 'of inestimable advantage. I stand upon higher ground than I did before', and before sending another letter to his parents in Bath he learnt that he was appointed second in command to Captain John Ross and was to be one part of a four ship expedition using three naval and one hired whaling ship. Two would be commanded by Captain David Buchan with Lieutenant John Franklin as his second, while he and Ross took the other pair – the whaler, and a very old brig. Preparations were to be completed at Deptford naval shipyard in time for departure early in May 1818.

After being away for so long he regretted the short time he had spent in Bath, particularly as Caleb was so badly disabled by his stroke, but he admitted he was enjoying London with a busy social life that was 'more pleasant to me than before; I now like it in gratitude for the advantages I am deriving for living in it'. He spent long hours writing to anyone who could supply any useful advice and information, often using letters of introduction provided by Caleb. These led to several exchanges of letters and personal visits to the astronomer Sir Thomas Herschel living near Slough. William Scoresby Jnr helped him with maps, drawings of the kinds of ice he would find and lessons about Arctic meteorology, while Sir George Hope, another member of the Admiralty Board, discussed the scientific work and the best methods for recording geomagnetic strength, and whom Parry described as, '*the* man at the Admiralty'.[12]

He used much of his time searching for everything of possible use being, as he said, 'on the alert'. Parry had a passion for detail and always prepared for his voyages carefully, a factor which considerably improved his chances of success. He considered new ideas and adaptations of practices used by the earlier explorers including Baffin, Bylot and Davies. Part of his reasoning was also that to succeed, welfare, safety, and health of his crew was paramount, and anything that mitigated the worst effects of sub-zero temperatures in near darkness was worth trying. He analysed each of the scientific objectives for his expeditions and the priorities they dictated. The instruments also had to be tested, so precise methods were needed for their evaluation and proper use.

In a letter home he expressed disappointment that John Ross, the expedition leader, would have all the scientific instruments on his ship.[13] He being allowed only three chronometers, a sextant, a dipping needle and pendulum – useful for navigation, but far too few for the research and instrument evaluation he wanted to complete himself. To make matters even worse, he was ordered to purchase a suitable barometer from his own pay!

The fourth person important as both mentor and friend was Sir Acheson Murray Maxwell (cf. 1771–1831). After a distinguished naval career under Sir Samuel Hood in the Mediterranean between 1805 and 1812, and surveying the entire coastline of the West Korean Peninsula, he had survived shipwreck, a statutory court martial, and was then knighted in 1818 becoming a Fellow of the Royal Society in February of that year.

During the short periods when Parry was back in London between voyages he visited Maxwell at his Pimlico home for dinner. Parry respected his opinions greatly and took his advice over officers, such as Henry Hoppner who had served with Maxwell in the Far East and became a trusted subordinate on three of his expeditions.

Maxwell gave wise advice on the politics, shifts in policy and current relationships between the two Secretaries to the Board of Admiralty, John Barrow and Thomas Croker, as well as its other members.[14] During these dinners Parry also asked Maxwell to transmit his ideas and wishes back to John Barrow and Melville prior to leaving on his expeditions.[15] Later, when living in Australia, he wrote to his sisters that, 'I feel it more incumbent on me than ever to correspond with Mr Maxwell'.[16] John Franklin also had a mentor, the MP John Whittington who lobbied the Admiralty Board members on his behalf, but never as successfully. He failed to persuade Melville or Barrow to provide Franklin the opportunity to lead the expedition he most desired, to find the North Pole and chart the Arctic Sea.[17]

The fifth person to play an influential role in his career was the combative, headstrong, ingenious but ultimately self-destructive, John Ross (1777–1856) (See Fig. 7). After a first meeting Parry said he admired Ross whom he described as, 'clever in a surveying way, and a good tempered affable man'. What he discovered later during their voyage however, was the impulsive, autocratic and irascible character of the man. An individual with little intention of attributing any success or honest recognition to his junior officers, including Parry, and not given to taking advice from subordinates however well informed and practical this was.

A red-haired heavily built Scotsman, his father had been a Church of Scotland minister, and was a veteran naval campaigner. He joined the Navy aged 9 as a volunteer, just as Parry had done. Under Sir George Hope in the Baltic

Squadron he had fought bravely and had first-hand experience with navigation in ice conditions. The two men had an acrimonious relationship for many years although Parry, in letters written to Charles, said that he would not deny him friendship despite all their differences if he apologised for his part in their original dispute.

But Ross could never bring himself to do this and through ill-judged criticism and hostility, as well his strong selfish streak, made his fellow officers wary and resentful for the blame he laid on others arising mostly from his own shortcomings. The first occasion being his account of the 1818 voyage with Parry which deeply annoyed John Barrow, antagonised key members of the Royal Society, and even threatened the authority of the Admiralty Board over their plans for polar exploration. His written account earned him £1,000 (worth £68k today) from the publishers but so badly undermined his own credibility that he was never given a naval command again. Even the scientific results from this pioneering new expedition were doubted and the British press, especially the *Times*, inclined to be sceptical about the Arctic project and finding a north-west passage, took any opportunity to satirise both the objectives and the results. He provided unhelpful reasons in the minds of the public to doubt its usefulness and the taxpayers any expensive participation. Ultimately Ross' disgrace and the official rebukes he earned for his multiple failures left Parry unscathed and regarded by Melville, Banks and Barrow as a competent officer whom they could rely upon. During the voyage with Ross he had neither neglected nor limited any tests of his few instruments, whatever the weather or other difficulties. Ross once ordered him to abandon some of these tests for lack of time but he had refused, stating that this was no reason for limiting completion of the experiments specified under their orders. Parry understood that in future his voyage accounts and especially any illustrations, must be confined to key subject matter using a proscribed style which was necessarily to be, 'plain distinct'. 'Sensationalism in the representation of landscape, subjects of scientific discovery, unusual flora and fauna, and dramatic events such as rubbing noses with Inuit people, would be unacceptable'.

The dangers from ice were so unusual that experienced whaling men or 'ice masters' were enlisted for every expedition and Parry became close friends with several of them. Men such as Mr Alison and Mr Reid. From them he learnt identification of the different types of sea ice, by their shapes and noises and the various risks they presented. His special interest however, together with that of Colonel Richard Sabine, an army surveyor co-opted by the Royal Society, was the measurement of magnetic strength in the Arctic. Extensive research was essential for precise measurements of the direction of the magnetic field which in the Arctic was expected to be especially strong – and moving!

His first expedition also showed him the qualities in both officers and men most essential for polar exploration. These being patience, physical stamina, tolerance of extreme cold, and the ability to withstand long periods of inactivity. British response to Ross' lurid voyage account also taught him a lesson in how the British popular press would exploit and satirise any lapses of judgement or comical references. He was not tempted to provide stories and memories to newspapers during his Arctic career although he advised Henry Barker and Robert Burford on two different polar panoramas but Burford found him too exacting and too occupied to attend to their own productions.

Public opinion was initially very favourable and excited over the idea of new exploration in the polar regions. The voyages of James Cook in the South Pacific were by now distant but still famous memories for Great Britain. His remarkable voyages and discoveries laid the foundations for the new generation of explorers and his second voyage in 1772–74 had been used as a basis for theatre productions including one by the brilliant Phillip de Loutherbourg, who depicted scenes from William Hodge's paintings of icebergs. One acclaimed production at Covent Garden Theatre portrayed a giant moving iceberg made of lathe and papier mâché, an early forerunner of the Burford and Barker panoramas built inside specially designed buildings during the early nineteenth century, the precursors to the modern cinema.[19]

The unknown and empty wastes of the far north were a romantic, fantastical and mystical blank canvas onto which writers, poets and artists projected their visions and plots from both a contemporary and an historical perspective. There was acceptance of a special sublimity affecting the Arctic region with its purity and separation from the rest of the world being especially attractive to people whose admiration of high mountains, large waterfalls, glistening glaciers and vast icebergs was spiritual and emotional. Francis Spufford quotes Mary Shelley who said, 'it ever presents itself to my imagination as the region of beauty and delights'.[18] Over and above this imaginative output including Shelley, Austen and Keats a sense of its important 'moral' attainment arising from new exploration epitomised the enduring human quest for both idealism and spiritual enlightenment. Joseph Conrad summed up this state by remarking that as a boy he read of 'the storm romance of polar exploration', and of 'idealists whose aims were certainly as pure as the air of those high latitudes where not a few laid down their lives for the advancement of geography; and of men of science it is difficult to speak of without admirative emotion.'[20]

These moralising, romantic and genuinely emotional *tropes* were also matched by other bleaker and more sinister associations including arctic demons, giants and ice monsters. Overall these contributed to the high degree of public interest in the departure of four small ships beginning once more the search for a north-west passage.

Chapter 4

A Pleasure Cruise

The correspondence between Parry and his family during December 1817 explains that the four expeditionary ships must be ready for commissioning by the end of January 1818. For this reason he may not return to Bath before leaving Britain. His health, he claims, is excellent but he is worried about his father's condition, the treatment he is receiving and why no recent reports have reached him.[1] It is difficult not to criticise Parry for failing to make one short visit to his home before leaving for an indefinite time but he did not do so.

Ross and Parry had met Sir Byam Martin at Deptford and two ships were chosen from the four being fitted out. They were all hired by the Government and neither the *Isabella* (385 tons) a whaler, nor the *Alexander* (252 tons) a small brigantine, were built for this kind of expedition. Naval shipwrights were ordered to brace and strengthen the bows on both ships and to fit bunks for the crew in place of the usual hammocks. Shortage of reliable information about weather conditions, monthly temperatures and the basic supplies needed for the Arctic is apparent from their inventories. The clothing provided consisted of 'milled' mittens, monkey jackets, ankle 'shoes', Flushing trousers, red shirts and swan skin drawers! Insufficiently waterproof or thermally protective for winter months in the severe conditions above the Arctic circle. They were clothes worn by most whaling crews during their purely summer fishing expeditions in the sub-Arctic.

Sixty wolf-skin blankets were loaded on each ship, together with rifles, a seven-barrelled gun and 3,600 rounds of ammunition, quantities of powder and shot sufficient for three years, while for trading with the Inuit each expedition ship had forty umbrellas, many sets of butcher's knives, looking glasses, scissors, axes, and brass kettles. The items closely resembling the inventory supplied to Cook on the voyages to the South Pacific fifty years previously!

Ross made known his dislike of both the two ships in a typically outspoken manner. He was told that he had no other choices and if he did not want them, he would be replaced! The Navy Board as parsimonious as ever had neither the money nor desire to build or convert more suitable vessels for Arctic exploration. For this voyage they had no intention of spending more money than was strictly necessary, nor to listen to the professional advice from a man such as Scoresby.

Ross and Parry found no consolation in the fact that the *Dorothea* and *Trent* – the two ships at Deptford which they had rejected, now being prepared for the Spitzbergen expedition – were even less fit for purpose. The *Trent*, Hooper noted, had to run her pumps continuously even in the Thames! It occurred to both men that preparations were being made hurriedly before Parliament could change its mind. Ross, Parry, Buchan and Franklin had to compile the navigational and limited meteorological advice needed themselves and to argue for any extra funds for items they wanted for the crews.

A list of stores taken on board included foodstuffs that were considered right for good levels of health, would promote physical strength, had above average storage potential, and were popular among the seamen. Staples included 1,300 lbs of salted beef, 18,200lbs of bread, 3,900 lbs of sugar and quantities of spirits totalling 1,925 gallons. Along with them went oatmeal, stocks of mint, savoury, celery seed, pickled walnuts and vinegar stored in iron casks.

Scurvy was the most serious health risk, although mainly due to Captain Cook's experiments the previous century and the proven use during the Channel blockade, their list of antiscorbutics included 2,275 lbs of lemon juice. Tinned food was tried out for the first time for both variety of diet and as an antiscorbutic, and thereafter became a normal part of Arctic provisioning. It was only a minor disadvantage that no tin opener had yet been invented.

Throughout January 1818 Parry was down at Deptford overseeing work on his ship the *Alexander*, but he had time to visit Banks' library and read all six volumes of Churchill's *Collection of Voyages*, as well as meeting William Scoresby Jnr. 'A very intelligent and indeed scientific man'. He was extremely occupied finding his crew of thirty-seven seamen, two midshipmen and his ice master; the *Isabella* needed fifty-six men and three officers. No lieutenant was to be needed for *Alexander* as Ross insisted that his ship would lead the expedition with the *Alexander* stationed two cables to windward of the *Isabella* and always within hailing distance to enable easy and swift communication. The instruction proved impractical due to the *Alexander*'s smaller size, was consequently considerably slower and much less manoeuvrable than *Isabella*. His order proved almost impossible to follow and Parry felt that the arrangement put him in the position of being 'a dog on a lead'. His experience and natural temperament were offended by the constraint and it increasingly annoyed him as the voyage progressed.

Other expedition members taken on *Isabella* were a young Greenlander, John Sacheuse, to act as an interpreter; Captain Sabine, an artillery officer and experienced land surveyor appointed by the Royal Society, along with two assistants. Their part was to try to establish the shape of the earth using Kater's

Pendulum, a newly invented device that needed frequent measurements taken at different points of longitude and latitude.

They were ordered to record at different times of day and night the position of stars and nebulae, especially the Aurora Borealis, and to test the seven different chronometers and new types of compass for comparable reliability and accuracy. Only Sabine was allowed to wind up the chronometers and was asked, since there was no voyage artist appointed, to prepare sketches and notes on subjects of botanical or ornithological interest. Being an army engineer and surveyor he had not been especially trained for this and although he tried hard he was later accused by Ross of inaccuracy.

Ross and Parry, as well as Sabine, were engrossed with calculating the magnetic forces affecting ship's compasses. This was chiefly conducted by Parry, who shared his results with Sabine and whose greater expertise developed and refined his methods.[2] Sabine was the protégé of Sir Henry Kater FRS, who was coordinating work on two related projects, the calculation of the exact size and shape of the earth and an isodonic chart of the world. (See Appendix 1. Geomagnetism)

Three whaling men were employed as ice masters and pilots. Mr Alison came with *Alexander*, and two others went with Ross. The men had years of experience in ice covered seas with prevailing Arctic weather conditions, and could interpret most of the atmospheric effects produced by different weather conditions over the ice pack as well as the likely consequences. (See Appendix 4. Meteorology).

Many visitors came down to Deptford as preparations continued, including Sir Benjamin Hobhouse, the press, senior officials from the Board of Admiralty including Lord Melville, Sir Byam Martin and Sir George Hope, and others from the Royal Society.[3] Neither Sir Joseph, now confined to a wheelchair, nor William Scoresby Jnr, who had been snubbed by Barrow on account of his attempt to persuade both Melville and Banks that he should lead this expedition, appeared. In March Parry heard that his elder sister Emma, married to barrister John Eardley-Wilmot, had died from consumption.

He was deeply affected, but so fully absorbed with his preparations that he did not visit Bath for her funeral, only writing to his parents:

> After the most busy day that I think I ever passed in my life, how happy I am to be able to sit down quietly in my lodgings, to attempt to answer the many anxious enquiries you have made recently respecting our expedition.
>
> I consider it to be our business to collect materials in as perfect state as possible, for the examination of scientific men when we return. I shall let nothing escape me that comes within my reach and I hope to be able to produce, on our return a tolerable collection for the learned to work upon.[4]

He did not allow his personal feelings to interfere with a vigorous attention to all his duties. The four ships finally left the Thames Estuary in April 1818.

The Ross expedition was ordered by the Admiralty to explore and chart sections of the north-eastern part of Baffin Bay, and then to follow any southern and western coastline. They were to start on the West Greenland coast and proceed northwards taking soundings and preparing coastal recognition drawings and surveys. These orders raised several questions and some doubts in both Ross and Parry's minds, since the time required for such a detailed inspection of the Greenland coast before cruising to the farthest northern extent of Baffin's Bay would limit the time left to look for any navigable route to the Northwest Passage. The exploration season finished in late September as winter ice started to form.

One earlier search for the passage led by Sir Christopher Middleton in 1741 had been unsuccessful.[5] He was then followed by William Moor and Francis Smith, who aroused suspicion among the directors of the Hudson's Bay Company who thought their expedition was a poorly disguised attempt to establish a rival trading business. The results were of little use other than to confirm that the southern-half of Hudson's Bay had no outlet on its western shore suitable for large ships. However, the northern side of this bay had not been examined since the voyages of Baffin and Bylot in 1616.

William Baffin had explored the northern side of Hudson's Bay and into the Foxe Channel in 1615, and charted a large area of sea between West Greenland and the coast of Labrador above Davis Strait. This was named Baffin Bay. During this voyage he had found three large inlets lying on its northern and western sides; the Smith, Jones and Lancaster Sounds named after the three sponsors of his voyage. He made charts with tidal measurements including a light current coming from the north-west into Davis Strait seeming to prove a channel existed to the west. His written notes were not as precise as those prepared by Cook when he charted the coast of Nova Scotia but despite this Ross took Baffin's commentaries literally, whereas Parry did not.

One totally speculative map was that of the Flemish cartographer Gerardus Mercator's Polus Arcticus which showed the North Pole as located on a large rock or island surrounded by four other larger tracts of land and a wide channel, the Straits of Anian, leading over northern Canada into the Pacific Ocean. It had no value for nineteenth-century explorers, but for others including John Barrow was still treated as evidence for the probable existence of the Passage. (See Map. 1b).

Once separated from the Greenland coast and its scattered Danish settlements, the Ross expedition had no way of communication with the outside world except by letters sent back using whaling ships as couriers. If they were to meet any. Parry was ordered that he must drop messages in six different languages in special bottles every day at noon giving their estimated position, these being one elementary precaution should the two ships become separated and a search instigated.

Their small wooden ships could easily be crushed by ice, sunk by a sudden storm, or rammed by even small icebergs. There was almost no hope of any rescue, especially during the winter which normally lasted seven months.

The European whaling fleets returned in October and came back in June. Conditions were volatile and weather forecasting reliant on those brief accounts by Middleton, Baffin and Bylot and their own whaling master's experience.[6] The voyage would seem suicidal today with risks as great as the NASA manned mission to the Moon in the 1960s. This voyage was intended, as was the space race, to be the only response to Russian competition and to prove beyond doubt that American science and technology led the world. In both cases it also offered opportunities to find new raw materials and determine how best humans could survive in extreme conditions.. The costs in both were huge and became disproportionate to the results actually achieved.

Hubristic determination to be first to find the North Pole and map the Northwest Passage had led Parliament to enact new legislation. In 1775 the Longitude Act was passed which offered a prize of £5,000 to any ship crossing the 110th meridian of longitude west and reached latitude 81 deg. north. Extra payments would be given the closer to the Pole or nearer to the completed passage anyone reached. Parry was to benefit from the first of these rewards, but not on his first expedition with Ross.

Parry's departing letter to his family written in the Davis Strait off Greenland was sent back with a whaling ship.[7] He confidently predicted that 'there is no doubt that we will get much further than any Europeans have done before,' and believed that the ice was rapidly disappearing but that 'from its appearance at one moment you cannot tell how it will be in ten minutes time'. In this eager and hubristic mood he continued by stating that it is the science rather than the exploration which lies behind this mission, 'the dangers of science on which I am engaged, measured by our short sightedness, are in reality nothing'. To console his parents, especially his mother, he finishes his letter encouragingly, 'we have every comfort and much more than I had before'.

As well as this news for family and friends, Parry decided to send John Barrow his notes on the variations shown by his compasses in tandem with the dipping needle. (see Fig. 20 and Appendix 1.) With the numbers he collected he suggested the magnetic pole was due west, and not in the same latitude as the geographic Pole. Ross' consent and knowledge over the despatch was not sought although his news was of significance, but it was likely that Ross would have refused permission or else claimed the discovery as his own. Parry possibly realised this and deliberately put himself up higher in Barrow's estimation, proving his analytical ability using the patterns of magnetic force.[8]

The letter was followed by another, accompanying a box of semi-preserved dead birds with a letter for Sir Joseph Banks, and a small collection of mineral samples for his father, apologising to him for there being, 'probably nothing especially new'. He wrote to John Eardley-Wilmot, his brother-in-law, strangely without offering any other condolences on the death of Emma, his eldest sister. His letter is about the rifle Eardley had given him, although this had 'not yet been used to shoot a bear'. The first sight of icebergs are described with enthusiasm and joy, but he shows impatience to get on and sail west into the unknown.

They reached West Greenland in mid-June of 1818 but were halted by a long barrier of thick sea ice created by the convergence of the Greenland and polar tidal streams named the Middle Ice by whalers. This delay they shared with a dozen British whaling ships. Given that the news in Britain was to expect a 'clearer sea, milder climate and dispersion of the polar ice,' this obstacle was unexpected and discouraging.[9] Several attempts were made to sail through the barrier but with no success, so they remained in Coguins Sound admiring the views and the constant stream of large sculpted icebergs drifting southwards.[10]

Captain Muirhead, one of the whaling captains, advised Ross to sail due north, hugging the west coast of Greenland, regularly checking the ice barrier for a break or 'lead', but the pack ice was 6ft thick in most places and crowded with icebergs. Their passage was slow and frustrating and at one time both ships became entangled by their anchors and nearly destroyed. A journey of just over 600 miles lasted over a month. When they reached Cary Island at the entrance to Smith Sound, they were greeted by a local Inuit tribe from *Etah*, a summer settlement, led by their chief *Tulloowah*, and were now entering an unknown region. The Scottish part of Ross caused him to name the place the 'Arctic Highlands', and to write an account of a meeting with the chief in a comedic manner. He had decided that both he and Parry would wear full dress uniform complete with cocked hats, medals and swords, and to follow the local custom of pulling noses as a way of greeting. John Sacheuse, the Greenlander and interpreter, made sketches. Several of these sketches and the report of several later meetings on *Isabella* with the 'Arctic Highlanders' were used in the voyage account but, due to both the superficial style and content and the Inuit pulling noses as their form of greeting, it only produced a satirical response in Britain. (See Fig. 9)

During the next few days they were introduced to a local Inuit woman who drew for them a map including conspicuous landmarks. This showed that the coastline ahead turned west and with this information Ross decided that they had almost reached the northern limit of Baffins Bay and it would be possible to sail south west, keeping close inshore and around the top of the ice pack.

They sailed past Wolstenholme and Whale Sounds along the north-western coast of Greenland where, at Cape York, they saw cliffs 1,000ft in height with patches of a strange bright crimson colour. At first they wondered if this snow was stained by bird droppings and Ross sketched a picture while Sabine was asked to suggest a possible scientific explanation. Bauer, a naturalist, later identified it as *Genus Oredo*, a rare type of lichen.

At the mouth of Smith Sound on the east side of Ellesmere Island they saw the 'beautiful glacier' at Cape Dudley Digger from which new icebergs were calving, some with extraordinary fissures full of colour and some resembling angels, birds, and pieces of classical architecture. Landing on one berg Parry estimated it to be 4,169 yards long, 3,869 yards wide, 51ft high and hard aground in 366ft of water. A huge polar bear was killed weighing 1,131 lbs and taken back to the ships for skinning.

Ross was unable to find a route or lead through this ice at the entrance to the Sound and rapidly decided that it did not merit further inspection. A premature decision he was to repeat a short time afterwards with serious consequences for himself and the expedition.

It was now August and the summer season for exploration was nearly over. Reports by Parry in his ship's log from mid-July onwards express irritation over both the lack of progress and Ross' autocratic behaviour, which was also corroborated by Hooper in his private journal. One instance occurred when Parry, measuring the magnetic variation with *Alexander*'s dipping needle, wished to compare his records with those of Captain Sabine on *Isabella*. He sent over a formal written message to Ross, 'I feel so much interest in the extraordinary deviation that if I can be of any assistance in your experiments on board *Isabella* tomorrow I shall feel a great deal of pleasure in taking part in them.'[11] He waited hopefully but got no reply and it is doubtful whether Ross even bothered to show his request to Sabine. This offhand treatment led to a deterioration in their relationship and for years afterwards Parry, conscientious and committed to the results they were collecting, affirmed that Ross did not always take the scientific tasks on their voyage as seriously as he should.

On 16 August they reached the end of the bay, passing more glaciers and numerous icebergs. Ross writes that he believes, together with 'all the different officers [I have], that this whole coast is connected all round. There is no navigable route to the north out into any Arctic Sea.' This opinion was delivered without discussion and without going fully into either Smith or Jones' Sound. He then ordered both ships to turn south and meet at the entrance to the remaining 'inlet' which was the position given by Baffin for Sir James Lancaster's Sound, the most promising inlet they had been ordered to inspect. Their orders had suggested that if the tidal stream measured lower down the Davis Strait was strong enough it showed that 'Baffins Bay *cannot* be entirely bounded by land as our charts represent it, but does connect, from the evidence of the currents, with the Arctic Ocean.' The weeks spent going north, investigating Barrow's theory that they would find an entrance to an Arctic Sea, had reduced the amount of time for exploration on the western coast of Baffin Bay. Only six weeks remained before they had to return home unless an entrance to the passage was found.

Ross' actions over the ensuing days led to recriminations, accusations and counter accusations between Ross and his brother officers, including Parry. These accusations continued between Ross and Barrow in a long running dispute that lasted over fifteen years. Ross defended himself by blaming all the delays caused by ice conditions and the poor weather contrary to information they had been given at the start, and repeatedly affirmed that he had diligently followed his orders, 'to investigate in entirety *both* sides of Baffin Bay, but, if no evidence of a channel was found, must leave the ice about the 15 or 20 September and NO later than the 1st of October.'[12] The word 'entirety' had been utmost in his mind. There were only a few weeks to complete their orders.

If they did not do so they would return home with their mission both incomplete and unsuccessful. Partly because of Baffin's remarks about the western side of his bay, and partly because Ross believed those remarks and could not see any exit in that direction, he made up his mind in advance that it was the case. But Baffin had never written off this possibility completely and Parry, reading the notes more carefully, decided that Baffin had really chosen to leave the question unanswered.

The account written by Ross said that the *Alexander* was sailing very badly and could not keep up with *Isabella* and was eight miles astern; according to others, Ross did not wait for *Alexander* and 'sailed away to the west towards a ridge of high mountains under full sail'. Lancaster Sound, or Inlet as Ross now named it, had 'land at the far end which I saw was a ridge of high mountains ... being extremely high in the centre'. After eighty miles, but still a long way from the end of the inlet, he promptly decided that there was no exit even though the weather was thick and misty. Without asking the opinion of his fellow officers on *Isabella*, including Sabine, he turned his ship about and sailed away from what he decided to call the Croker Mountains after the First Secretary to the Admiralty. Parry, trying to keep up with *Isabella*, then met her going past him in the opposite direction and was bound to follow behind. Ross made no attempt to reduce sail so that *Alexander* could come close up to let him explain to his second in command his reasons for sailing away.

The Croker Mountains did not exist. Their appearance and size were a mirage created by low sea temperatures and sea ice which refracted light up onto clouds. On this occasion they resembled from a distance substantial mountains, permanent and impenetrable. Closer inspection, despite the worsening conditions, would have shown them not to be so, but the miles of unexplored coastline to the south of his position and with two weeks of cruising time remaining, Ross made the over hasty but not entirely unreasonable decision to turn round.

His worst failure was not to explain his decision or consult with his fellow officers, especially Sabine and Parry. They might have convinced him that the mountains were a mirage, since on topographical detail as large as this, Baffin's statements would have been unequivocal. Sabine's opinions were founded on his knowledge of meteorology and conversations with Scoresby who had warned him to be on his guard against such curious effects.[13]

Ross argued that from daily measurements no tidal stream could be found issuing out of Lancaster Sound; he suggested that if a measurable tidal current was found lower down in Davis Strait, it had to be flowing from a channel to the south-east of his position. Despite this sound reasoning the case held against him was that he had not followed an important part of his orders which were to 'fully' ascertain any potential route to the west. After waiting weeks to move from West Greenland to the North American shore and with both ships in a reasonable state of repair and provisions sufficient for a longer stay, his premature return seemed like expediency rather than a carefully considered decision, especially since it was taken without canvassing the opinion of any other officer.

Over the following two weeks they sailed south and east, discovered and named Bathurst Bay and completed the circumnavigation of Baffins Bay. They continued measuring magnetic variations, deciding that the North American coastal variations were stronger than those on the West Greenland side. This Parry wrote, made 'their binnacle compasses virtually useless'. Their tables were invaluable for later voyages and for James Clark Ross, John Ross' nephew, one of two midshipmen on the *Alexander*, they helped identify the position of the Magnetic Pole thirteen years later.

On 25 September Ross wrote that winter had arrived and all the men were issued with extra clothing. A week later they set course for England during which they survived the worst storm of the voyage, lasting eighteen hours. It was so severe that Hooper, a religiously inclined man, wrote in his private journal, 'may we never forget who it is that every trial and every danger is our Guide, Protector and Friend'.[14] Most knew they were very fortunate to survive.

They anchored off Orkney on 11 November 1818 and the recriminations started almost at once. It was Hooper who expressed the mood of many of *Alexander*'s crew that:

thus vanished our golden dreams, our brilliant hopes, our high expectations. To describe our mortification and disappointment would be impossible at thus having our increasing hopes annihilated in a moment, without a shadow of reasoning appearing; we have returned without adding one iota to the discoveries of Baffin two hundred and two years before.[15]

Captain Sabine was next to contribute because of a dispute with Ross, who had casually misattributed some of Sabine's scientific work to himself. Ross at first refused to acknowledge this failing and then tried to dismiss Sabine's work as 'untrustworthy'. Ross' nephew James, a midshipman on the *Alexander*, weighed in by contradicting his uncle publicly, declaring that their premature return lacked a proper reason. Parry alone among the officers kept silent except in letters to his family and his private journal. He felt Ross was at fault and had acted on a whim. Ross had tried to exonerate himself, declaring that he had involved crew members of the *Isabella* in his decision, but everyone else denied this. Parry realised that Ross, as he put it later, 'was beyond argument'. Loyalty to his commanding officer was no doubt in his mind and open conflict with a superior officer might damage his own prospects, so, wisely, he decided to keep quiet knowing that 'blundering Ross' would seal his own fate.

It would only be fair to acknowledge that Ross complied with Admiralty instructions to investigate the 'entirety' of Baffin Bay, and still adhere to the time limit that had been imposed. It was also unfortunate for Ross that John Barrow had been badly disappointed with the result of the other expedition to reach the North Pole via Spitzbergen and considered Ross' actions even more reprehensible.

The anticipated 'contraction' of ice described by Scoresby was never evident and the Buchan and Franklin expedition was prevented from going further than the northern tip of Spitzbergen or Svalbard. They were forced into pack ice under storm conditions and badly damaged. The ships struggled slowly back to Britain with little to show for their efforts apart from some coastal soundings, surveying onshore and wildlife sketches by Lieutenant Frederick Beechey. Their voyage had been so unsuccessful that Commander Buchan did not attempt to write an account of the voyage and John Franklin suggested to his friends it was ultimately 'pointless'. He was clearly anxious not to antagonise Barrow, hoping to be chosen for another expedition this time with overall command, but Parry, now with first hand experience of the complex ice formations in Baffins Bay and its abnormal meteorological side effects, had the full support of Sabine. In addition, he had his daily records for the variations and corrections necessary to adjust compass readings in all parts of Baffins Bay.

John Franklin now proposed an expedition down the Coppermine River in small boats following a similar route taken by Samuel Hearne's expedition in 1770, and to explore the middle part of Canada's long polar coastline. Parry meanwhile was being considered for the command of another expedition to Baffins Bay and to re-enter Lancaster Sound or Inlet and search for an outlet to the west.

While the Admiralty, the Royal Society, Melville and Barrow debated both the different plans, disputes over the Ross expedition went on undiminished.

Viscount Melville decided to hold a Parliamentary enquiry into John Ross' actions, while Ross succeeded in upsetting not only Croker and Barrow but several other Admiralty men by suggesting he was given a court martial which, he felt, would exonerate him, but which he surely realised would have embarrassed many others. He also applied to lead the next expedition to Baffins Bay but was two weeks behind Parry in doing so. Maxwell had advised Parry to write immediately to Melville and Barrow telling them that a route existed which was precisely what Barrow wanted most to hear.

To add to his many shortcomings, Ross published his own version of the voyage using a series of coloured engravings, but without obtaining Admiralty approval or bothering to show the draft to his fellow officers. Barrow was infuriated by this defiant act, especially as sceptics among the British press considered the failed attempts on both the Pole and the Passage were pointless, and now possessed excellent material with which to ridicule and criticise the results.

Ross' account, dedicated to Viscount Melville, was published in March 1819. The engravings chosen by Ross were based mainly on his own sketches.[16] They included the meeting with his 'Arctic Highlanders', pulling noses, crimson snow, large polar bears diving off icebergs, the peaks and glaciers around Cape Dudley Digger, and distant views of the Croker Mountains. The latter sketches he fiercely maintained were made at the time but, suspiciously, given the speed with which *Isabella* turned round, are almost certainly based on memory and for self-justification purposes. The captions to the engravings included the use of words such as 'remarkable', 'peculiar', 'towering' (for icebergs) and 'dominating' for mountains and coastal features. His drawings were inferior to those of his midshipman Henry Hoppner, the officer chosen to draw events and discoveries during this voyage. Hoppner was chosen because he had 'a skill in drawing so considerable as to supersede the necessity of appointing a professional draughtsman.[16] Wishing to popularise both the journal and himself, Ross had produced a combination of exuberant descriptions, amateurish and garish illustrations including John Sacheuse's sketch at *Eliah*, and numerous claims of his successes which all provided good entertainment for the reader but much scepticism, outright derision and criticism among newspapers such as *The Times*.

Cartoonists such as George Cruikshank made full use of the various 'discoveries.' Ross on a sledge pulled by dogs, baskets of snow balls in different colours, men with missing noses which had been 'pulled off' by Eskimos, and large polar bears ridden by 'jolly jack' sailors. John Barrow was concerned that such satirical commentary could undermine parliamentary support for further expeditions. Two months after Ross' book appeared he retaliated with a scathing attack on his actions in the Royal Society's *Quarterly Review*. The voyage account was dismissed as nothing more than 'a few months voyage of pleasure – a voyage which any two of the Yacht Club would easily accomplish'. His conduct was 'impenetrably dull or intentionally perverse' and, most damningly of all, that the existence of a north-west passage was considerably strengthened by what he had *failed* to do. After this rebuke came a more damaging one in a leaflet published by Captain Sabine run in several newspapers and confirming that Ross had never tried to consult him or anyone else in Lancaster Sound about the Croker mountains, and his reasons for turning back. Ross denied everything but Sabine maintained that 'the Captain said he saw land when we were all "at dinner"', which meant that 'they had been seen by one individual, for a short time, at a very considerable distance, on a very unfavourable day'. Sabine's opinion was that this should be considered as 'uncorroborated evidence' and was 'not decisive'. Ross was discredited in the opinion of Barrow as well as most of the senior ranks at the Admiralty and was never commissioned again.

His pride, obstinacy and sheer tenacity caused him to return to the Lancaster 'Inlet', the source of his disgrace, years later with the single intention of salvaging his lost reputation and of exceeding the achievements of Parry.

Chapter 5

Through Lancaster Sound

By January 1820 Parry was immersed once again in arrangements and decisions for his next voyage. John Barrow wished to lose no time fitting out the next expedition which he hoped would prove to the public and parliament that Ross was mistaken and that Lancaster Sound contained at least one channel out to the west. He was determined to prove that the Admiralty and the Royal Society were right to continue the search for the passage, but to do that he needed a commander with conviction over the passage, experience of ice conditions, polar experience, navigational expertise, and above all unshakeable resolve to succeed whatever the obstacles. The starting point of this expedition was to be the western side of Baffins Bay, as yet unexplored. (See Map 1a). Shown on the 1794 map of Samuel Dunn as Prince Williams Land and Arctic Lands, it is an entirely blank space.

Parry's carefully judged letter to Barrow from Greenland with content establishing his work on compass variations in order to find *Isabella*'s and *Alexander*'s true position, had shown his capability with navigation in risky situations. He had had to land on icebergs to obtain clear readings with the dipping needle away from the magnetic influence from metal items on the ships. Sabine and Parry had both proved equal to the work and of recording accurate measurements. Barrow now trusted their joint opinions and capability to cooperate together over analysis of the many scientific trials needed. He also clearly welcomed their shared conviction over a Lancaster Sound route.

Ross immediately offered to be commander again as soon as he heard about the new expedition but his standing was not high and he continued to show his remarkable capability to offend his seniors, so that few members of the Board of Admiralty wished to appoint him again. Parry, they believed, was now the better and wiser choice, despite his young age of 29.

Parry met with Maxwell over dinner at his Pimlico house several times during the winter of 1819. Maxwell encouraged him to write a letter to Lord Melville expressing certainty over the existence of the Northwest Passage and all his capabilities for leading the expedition. He had suggested too that proposals should be given for fitting out the ships to be used.[1] Two letters to his parents refer to meetings and interviews with John Barrow and Sir George Cockburn whom he describes as the 'mainsprings' of this 'affair'. Both men questioned him closely about his proposed improvements to rigging, clothing and rations.

His letters home are brief due he says, to his hectic life in London visiting important Admiralty officials and the 'machinery' of the Navy Board; but are also affectionate and solicitous about his father's poor health, and ask questions about the family nephews and nieces belonging to his late sister Emma. They were all compelling reasons to visit Bath, but he did not.[2] It seems from the second letter that he hoped his younger sister, Caroline, would look after Emma's children and had written earlier to Charles along these lines before asking Eardley-Wilmot, who was completely against this idea. The suggestion was most likely driven by deep emotion and eagerness to help resolve family issues before leaving them all for a long period, absence making it impossible to contribute any further suggestions or express views about the future for his nieces and nephews. He had had doubts about Eardley-Wilmot as a husband for Emma, but following his brother-in-law's remarriage and future care of his six nieces and nephews a wary but friendly relationship developed. After a letter in which these and more general family matters were raised, he wrote again the next day telling them that the Arctic command was his.[3] Maxwell had already heard that both Barrow, Sir George Cockburn, as well as a Mr Jepping, were strongly in his favour. He was not promoted from commander to captain which he might have expected, but his attention was fully occupied on the preparations for his two expedition ships, HMS *Hecla* and HMS *Griper*. He began choosing his crews, ninety-one officers and men in total, as well as calculating the quantity of stores required and the maintenance requirements for both ships.

The excitement in this letter is evident, although he expresses disbelief that John Ross had been promoted to post captain despite the disputes with himself, all his officers and Barrow. But it now mattered far less, as he had secured appointment as the overall commander with better ships and the full confidence of the Admiralty behind him. *Hecla*, unlike the *Alexander*, 'will be as perfect for our purpose as human art can contrive. The interior arrangements, as also the masts, rigging etc. are left to myself.' *Hecla* was a bomb vessel designed to bombard ports with a heavy cannon or mortar. He used her for all his future polar expeditions having, he said, 'a strength, a capacity of hold, good sailing and fine accommodation'. The hull design with specially strengthened bows, was capable of breaking through moderate formations of ice and became the type of vessel used in both polar regions. The *Griper*, a small gun brig, was chosen by the Admiralty as consort but after complaints from Parry that she was 'utterly unfit for this service', she was refitted with new raised decking to make her crew quarters more spacious, and to allow for more storage space. Lieutenant Edward Hooper, the officer appointed by Parry as his expedition purser as well as a scientific observer, described the *Griper*'s sailing capability as 'lubberly', meaning slow and difficult to keep up into the wind. She became like the *Alexander*, a source of delay, which sadly contributed to several missed opportunities.

The officers Parry wanted included Lieutenant Frederick Beechey as second in command on *Hecla*, who had been on the abandoned Spitzbergen expedition, Alexander Fisher, the Assistant Surgeon previously on *Alexander* and an appointee of the Royal Society, John Edwards the Surgeon from *Isabella*, and Captain Edward Sabine, who would again lead the scientific experiments. Parry had lobbied the Royal Society to allow the release of the cheerful Captain Sabine, despite attempts by the Admiralty to prevent it on the spurious grounds of cost!

Lieutenant Matthew Liddon, an old friend of Parry's from the North American days on HMS *Maidstone*, was given command of the *Griper*. He also chose James Clark Ross, John Ross' nephew, as one of the five midshipmen taken who later became a renowned polar explorer and very close friend. Some of these choices seem to be the result of advice provided by Maxwell since he well understood the conflicting preferences for different individuals and their capabilities held by the Admiralty Board. One striking feature of this expedition was the youth of its naval officers – at 29 Parry was the oldest and even Sabine was only 30. More advice given by Maxwell was that he should acknowledge in writing their Lordships 'expectations' in advance of his formal written instructions, and to list all the supplies and equipment he must be allowed which, as Parry later admitted, were obtained on 'a biblical scale', despite their cost.

Shortly after the appointment Parry moved to number 3, Downing Street, at a rental of £1 a week. His new address was arranged by John Franklin through his London connections and the move away from Suffolk Street and its proximity to the Admiralty Offices in Somerset House was timely, as being at that address meant daily soliciting from people wanting to join his expedition.

To one man, a connection of Fanny Burney, he was polite yet firm and answered that despite his 'gladness to do anything for anybody connected to Miss Burney people such as a Lieutenant Spillers, an Army Surveyor and Engineer, cannot be admitted'.

A sympathy for Franklin is shown in letters to his parents since being junior to Franklin in years of service, he had nonetheless been the officer selected:

> Franklin is low spirited and uncomfortable. To make up for his disappointment the Navy have given him every hope that he will be sent next year on an expedition to the North. His conduct has been so very handsome to me that we must all sympathise with him in his disappointment. Sir George Cockburn and several others at the Admiralty have pushed for his promotion and I trust he may now get it.[4]

Years of disappointment and frustration during his time in the Channel Fleet, the Baltic and on tedious convoy duties, as well with the North American squadron waiting for a command, lay behind his very genuine sympathy. The admiration he felt for Franklin and his wartime record did not get in the way of his own ambitions – leadership and success where all others had so far failed.

Franklin generously showed no resentment over Parry's selection and none of the letters between them that exist give any sign of this, although remarks made much later by his second wife, Jane Griffin, imply that she felt he, Parry, was responsible for several of her husband's disappointments. If Franklin had gone he would have been in overall command for reasons of seniority and Parry, understandably, was unwilling to accept this.

The new proposal made by Franklin was accepted by John Barrow, Viscount Melville and the Colonial Office, and welcomed by the Royal Society, who helped contribute to its cost. It was an expedition comprising a dozen seamen three other officers and two naturalists to explore and map the long Coppermine River in North Western Canada and then survey the northern Arctic coastline seeking proof of a navigable passage from the landward side.

Work on the ships progressed smoothly with Parry refusing to make either compromises or to delegate any inspections. Nothing could be allowed on board which might impair the safety and efficiency he had set himself. He supervised the workmen at Deptford shipyard, inspected and tasted the recipes for their canned foods, chose suppliers of antiscorbutic lemon juice and vinegar, as well as all the concentrated alcohol needed. He specified and tested all the special clothing for the crew which he had helped design and the new stove for keeping the crew's quarters more comfortable at night and to help dry their clothing.

A few short letters to his parents give a running account of all this work which was combined with a busy social life visiting people such as Lords Lowther and Melville, Sir Joseph Banks, Henry Kater the astronomer, Sir Byam Martin, John Franklin and his friend and mentor Acheson Maxwell. He attended a course of weekly lectures at the Royal Society about minerology along with Sabine and Alexander Fisher, given by a Mrs Lowry. Interest in both geology and minerology was inherited from Caleb, his father, for whom he kept rock samples collected in the Arctic. Selecting and choosing the books they would need he asked Sabine's advice to avoid duplication and despite all these many commitments he found time to help his uncle Benjamin Hobhouse become an MP, as well as reluctantly agreeing to sit for his portrait painted by Sir William Beechey RA. This painting was paid for by Miss Burnett, a friend of the family, for the large fee of 150 guineas (worth £10,000 today), which he found embarrassing and did not like! (See Fig. 5).

As their date of departure approached his letters continue to be about the welfare of the six Eardley children, his younger sister Caroline's future, and his father's health. He was preparing them for a long absence and offered opinions over family matters including one that Caleb, 'must not be given turnips to eat since it is his stomach and not his head that occasion his attacks, especially as he is staying still for so long'.[5] His disappointment was that Eardley-Wilmot had not written to him or paid a visit, but he also admits to not having had time to visit or even write to many old family friends, including the Herschels, Baldwins and Wilsons, all of whom his mother asked him to contact.[6] The time needed to communicate with many family friends and relations, keep up his naval connections, and attend to all the inevitable last minute hitches, became very stressful and when eventually ready to depart on his expedition he felt deep relief and gratitude.

By early April the ships were ready to go downriver to Northfleet where the final fitting out would be completed, all the different instruments had to be loaded, and the men paid three months' advance wages from which they were ordered to buy their clothing. The *Griper* then went further downriver for sea trials and was deemed top heavy and unsafe so more ballast was loaded and some provisions taken off.[7] At this moment John Ross published an attack on Parry and Sabine in a new pamphlet. He described Sabine as not a true 'naturalist', while Parry's accuracy with compass corrections was questioned, but despite this provocation Parry decided not to respond since 'the new expedition is more important than continuing the controversy with a man like Ross'. Sabine, however, quickly produced a joint response titled: 'A refutation of certain misstatements in a book entitled "Narrative of a Voyage of Discovery" by Captain John Ross KS'. The pamphlet was printed and distributed a few days before leaving to ensure Ross could not accuse them of 'running away'. Barrow, on reading Sabine's pamphlet, said that had his facts been known immediately after their return, John Ross would have been court martialled.

Many visitors came to see them before they left, including Sir Henry Hotham, a hero of the Napoleonic Wars, Lord Melville, Lords Minto and Temple, Maxwell, Lady Sheffield, and his old commanding officer Captain Ricketts from the *Ville De Paris*. By this time Parry felt worn out guiding and explaining everything to everybody. A bad infection in his left arm which the ships Surgeon Fisher had to bleed made matters worse, but fortunately for him it healed fast. Such a difficulty only weeks later at sea could have become far more serious.

Among the visitors from Bristol was a Miss Browne, a niece of Edward Sabine and a relation of Lord and Lady Sheffield. Parry demonstrated to her group a special life preserver, a costly item given him by his parents. He placed it around her waist, tied it up and invited her to inflate it as part of the demonstration.

Two years later the consequences of this seemingly innocent act rebounded unhappily for Parry. The intimacy implicit in this harmless encounter with a young unmarried woman represented an attachment which could be taken by her friends and her mother as a mark of Parry's intentions, however unintended. It was probably an indication of his inexperience and naivety with women.

Parry wrote to Caleb and Sarah on 13 April just before going downriver, giving them the date for sailing as, 'any day in the first week of May'. This gives him a sense of deep relief and he uses stirring words about his feelings over the expedition, 'a great deal depends on me. We have everything we need and I almost long to be at work in the ice.'[8]

The Admiralty instructions handed to Parry at Northfleet were less contradictory and capable of misunderstanding than those given to Ross. He was to sail directly to Lancaster Sound and to use:

> [his] best endeavours to explore to the bottom of that Sound; or, in the event of it proving a strait opening out to the westward, to use all possible means, consistently with the safety of the two ships, to pass through it, and ascertain its direction and communications; and, if it should be found to connect itself with the northern sea, you are to make the best of your way to Bering's Strait.

Once through Davis Strait all decisions were his own but this principal objective was clear and unequivocal.

On reaching Greenland off Cape Farewell on 4 June, they celebrated King George III's birthday with Scotch reels and country dancing; as some fog cleared, they saw the first icebergs which lay around both ships. These made a strong impression on William Hooper and in his private journal he writes:[9]

> A scene which I feel inadequate to describe with justice and which, I believe, will be as difficult to conceive; it seemed to partake altogether of another – a better world. It was associated with the idea of all that is good and great and gracious. Of the power and skill of creation.

The sight of the Greenland ice pack brought out similar feelings from Parry even though he had seen this spectacle before. His journal records: 'The pure white of the snow upon the ice, the exquisite rich blue about the waterline and the numberless shapes which different masses of it assume, create in us fresh expressions of admiration whenever we look at it.'

His final letter home is of sadness blended with enthusiasm and excitement at their future prospects. The sadness was over his sister Emma's premature death, 'his blessed angel Emma – taken by God to himself', followed by: 'we see only the outside of things for we cannot comprehend the intricacies of God's whole system'.

Determined not to become trapped behind the middle ice on the western Greenland Coast as the year before, Parry decided to strike across Baffin Bay as soon as he saw a lead. The 1818 expedition had been forestalled by the northerly drift of ice round Cape Farewell driven by the Gulf stream. This array of icebergs then met more large icebergs coming south from glaciers in the north. It was a risky decision, especially as the pack ice down the middle of Davis Strait was very heavy that year, but despite the conditions they moved slowly forward by hauling the ships using their small boats and heaving on ropes anchored on the ice ahead, a process known as 'warping'. These efforts went on for days and were extremely arduous but they reached clear water on 28 July and were then able to set sail for Lancaster Sound. The sailors had pulled the ships eighty miles through dense ice pack, encouraged only with a special daily issue of spirits and changes of dry clothing. Their reward was a month's advantage over Ross' arrival at the same place one year previously.

On the *Hecla* and *Griper* opinions over whether the Croker Mountains were real or whether open sea lay ahead reached fever pitch. Lieutenant Hooper witnessed these emotions 'an agony of feeling and a day of fearful anxiety'. Fisher the surgeon described both sets of crew feverishly climbing the masts looking for 'what I hope we shall never see!' The tension increased until they reached the position where *Isabella* had turned back the previous year and saw no mountains on the horizon. Hooper wrote, 'joy lighted every countenance'. They had, he added, 'arrived in a sea which had never before been navigated, we were gazing on land that European eyes had never before beheld'.[10]

Chapter 6

'No Common Men'

The *Hecla* and *Griper* now entered a broad channel reaching out to the distant horizon bordered with high cliffs interrupted by gloomy chasms, wide inlets and narrow fjords. Above these rose the peaks of snow-covered mountains. One inlet forty miles across lay on their port side, the southern coastline of Lancaster Sound. The entrance was guarded by two massive headlands several hundred feet high, one standing at the northernmost end of Baffin Island, the other, Cape Clarence, being the tip of the North Somerset Peninsula, as it eventually became known. This inlet they named Prince Regent Sound and was to be the location seven years later of Parry's narrowest escape from disaster.

Broken but navigable sea ice lay ahead and they decided to explore another inlet which appeared to lead out towards the south west. After one hundred miles of cautious progress, sometimes in thick fog, the way became entirely blocked by ice and they were forced back up into Lancaster Sound. After these diversions Parry decided to keep sailing straight ahead and to resist any urge to explore other inlets – however promising they seemed. They passed through the western entrance into Lancaster Sound and then reached the Canadian polar archipelago, or polynya, which stretches across the extent of the central North American landmass; a wide expanse of ocean in which lay a confusing maze of islands, shoals and channels covering hundreds of square miles.

By keeping unswervingly to their westerly course and helped by a following breeze and little ice, the chances of emerging out into the Pacific Ocean before the summer season ended seemed excellent

This section they called 'Barrow Strait' following Admiralty instructions that all new discoveries were to be given indisputably British names for future cartographers. It was an established practice, used by Cook on all three Pacific voyages, by adopting names of royalty, senior naval officers, Admiralty officials and influential politicians, shuffled about, allocated and sometimes reallocated; a sort of tribute left for the individuals sponsoring the voyage. This was the procedure for any newly discovered land or conspicuous landmarks. Parry followed the convention giving the most important landmarks and geographical features to individuals who had supported his voyage. Even the first secretary of the Admiralty John Wilson Croker, whose 'mountains' had been found to be non-existent, was rewarded instead with a large bay on Devon Island, part of Lancaster Sound's northern coast.

The weather remained clear apart from a few snow showers so that fixing the positions of islands and landmarks progressed well. By 12 August, they had sailed over 300 miles from the start of Lancaster Sound and Parry next decided to discuss with his officers, the ice masters and Captain Sabine whether they should continue through Viscount Melville Sound, towards land just visible in the distance which they called Melville Island. The new landmass had cliffs and hills over 1,000ft tall and was 150 miles long and seventy-five miles wide.

The decision whether to continue needed to be taken because continuing further west would eliminate the prospect of returning home that same year, until and unless they reached the Pacific Ocean. The officers all agreed to continue and after a few days nearing Melville Island, their speed dropped and surface ice began to accumulate at night which then consolidated itself during the lower daytime temperatures.

Despite the obstruction and areas of fog they reached longitude 110 deg. west of Greenwich on 4 September. This position and its significance was announced to all the sailors as it gave every man a share in the parliamentary bounty of £5,000 (around £490,000 today). Their response, Parry noted, was 'joyous'. His share as commander was £1,000 (around £96,000 today). The bluff headland immediately to the north was named, appropriately, Cape Bounty; they anchored nearby, raised ensigns and pennants and celebrated in a place they named Hecla and Griper Bay. This, he admitted, 'gave him no ordinary feelings of pleasure to see the British flag waving for the first time in these regions, which had hitherto been considered beyond the limits of the habitable part of the world'.[1]

Landfall on the eastern end of Melville Island was reached on 6 September and although now surrounded by ice floes, Parry refused to wait longer. He decided to send several parties of men ashore to hunt for game and 'to make some observations'.[2] Two of these groups became lost in thick snow and were unable to find their way back. Even by these early autumn months there were still fourteen hours of daylight and four more of dim twilight. The men became disoriented in a snowstorm and search parties were sent to look for them. By a miracle the first group regained the ships luckily finding and following the footprints of one of the searching parties. The two groups who survived three freezing nights in the open were completely exhausted and some men severely frostbitten. Parry ordered that no man should leave the ships in future without five days' food, far more clothing, and carrying pikes with flags tied on to mark their route and their way back. Bottles with compass bearings for the ships were to be tied to each pike.

On the far distant western horizon the sea was covered with ice and a powerful gale began to blow out of the south west. Despite these conditions and because there was thinner ice around the ships they could sail on, beating into the wind.

In temperatures as low as 10°F (−12°C.) on 17 September they reached a longitude of 112 deg. west. Meanwhile, Franklin and his men were half-way down the Coppermine River in Northern Canada, and further to the east at 108 deg. west. The mouth of the Coppermine River was actually 5 deg. further west, but Franklin's maps had been over-corrected.

On 20 September the weather deteriorated badly and their different thermometers dropped below −15 deg C. so they returned to Hecla and Griper Bay. Parry now began to prepare for the long winter ahead, anticipating a six-month wait before spring arrived. He had prepared for this, having two years' stock of fuel, antiscorbutics, food and chandlery.

First a sheltered bay was needed on the south side of Melville Island and through the new ice they cut a canal to bring them nearer to the low beach. This position in shallower water providing better protection from large icebergs being driven eastwards out of the Polar Sea by strong winds.

The canal they constructed was hard work using special ice saws each handled by four men. The job needed to be completed fast since few could withstand spending all day in seawater in temperatures of 2 deg C.[3] Daylight hours were becoming ever shorter and the ice soon filled in places cut out the previous day. The sailors took to this arduous work enthusiastically, inventing ingenious ways of sailing sections of cut ice down the canal to keep the water clear for the ships. At the end of three days of supreme effort, with much encouragement from their officers, the task was finished. *Hecla* and *Griper* were hauled up the 2¾-mile long canal through seven inch thick ice and into their Winter Harbour, arriving at 3.15pm on 26 September amidst rousing cheers.

The expedition had completed an extraordinary 650-mile voyage through Lancaster Sound, down the Barrow Strait and had almost reached the end of Melville Sound.

This route was officially recognised as the Parry Channel in 1956 by the Canadian Board for Geographical Names.

They now started lowering the ships yards and topmasts, using them to create a covered deck area for exercise. Ovens were built in the main hatchways and pipes connected to the galley ovens to send hot air round both ships. Snow was mounded up on both sides of the ships to give more insulation and wind protection. Because Parrys' main concern was the physical and mental health of his men and how it could deteriorate during the long winter and because of his time serving on the Channel blockade, he had observed signs of listlessness among the sailors, many made unfit from chest and lung complaints. He therefore decided in advance to organise occupations for each watch. Successful completion of the passage during the next summer season depended upon his men being in good health and willing and confident they could finish the commission.

He had foreseen that the relationship between his officers and their men in this unusual situation needed to be different. A relationship that would be welcomed but not exploited by the men. He hoped the people he had selected would accept responsibility for their own and each other's welfare like an extended family. Daily routines were planned and shared out, of which several concerned the daily airing of bedding, monthly medical inspections, exercises on deck, and the time after evening prayers given over to music and dancing. He ordered the cooks to create new recipes with as many variations as they could, not only to help provide a balanced diet but also to provide interest and variety to meals. Other distractions using entertainments and regular Bible classes were introduced to reduce the insidious effects of boredom and depression caused by the dark winter days.

The officers took watches in the observatory and were ordered to report anything unusual however small or insignificant they might seem. He instilled in everyone that 'what is seemingly unimportant, may prove useful and interesting'. This principle was again repeated in his orders during his second expedition in 1821–1823.

The portable observatory used by Sabine and Fisher the astronomers, was erected after they had built a small forge to make pickaxes for holes into which supports needed sinking in the permafrost. Officers recorded hourly the weather conditions, temperatures, visibility, magnetic variation, astronomical position of stars, and all bird and animal sightings. This continued for twenty-four hours a day. The telescopes were found to be awkward to use because the eye pieces froze fast to the face and condensation from their breath froze, clouded over and then crystallised on the glass.

By 16 October daylight had almost disappeared completely, with the sun just above the horizon. Temperatures barely rose above freezing at midday and at night they showed -26 degC.

Plans for overwintering had been made based on winter conditions reported by whaling ships, but none of those had been at the same high latitude for so long. Scoresby's advice had been taken over clothing and foods but Parry knew that most of his decisions would be a matter of trial and error, and that survival was dependent on most being right. Hooper, Liddon, Fisher and Sabine were all consulted for their observations and opinions about conditions and any improvements that could be found.

One scheme to keep spirits high was the preparation of theatrical plays performed by officers along with a few willing sailors. Considerable ingenuity was employed in the making of scenery and for various props and costumes. A theatre committee managed by Lieutenant Beechey was created and the first play staged was *Miss in her Teens* by Garrick. Parry played the part of Fribble, and Hooper wrote, 'it was a resounding success'.

It was unusual for officers to take part in such activities but Parry enjoyed both acting and the theatre. Plays were rehearsed and performed every fortnight and after they had used up the stock of material from the ship's library, they began to write their own.[4] Another distraction was the ship's newspaper called the 'Winter Chronicle and North Georgia Gazette', with Sabine appointed editor. He was good at this work and being an independent voice, was willing to make light-hearted contributions some of which were audacious commentaries on naval life. Ideas were solicited by installing a special suggestion box at the foot of the companion ways and stories came in about the food, the cooks, the naming of unpopular dishes, the repetitious menus as well as some serious reviews of the plays. In a few editions, poetry composed by Parry was included.

After the Christmas feast a six mile walk was led by Parry, followed by an extra ration of rum, singing and dancing round a portable organ and music made by officers who played instruments.

January and February were exceptionally cold with deep snow covering both ships and extremely low daytime temperatures. Two of the chronometers in Parry's own cabin, the coldest part of the ship, stopped working, while the day cabin for the other officers was barely habitable. Everyone sat huddled in greatcoats and pretended to read, play chess, or practice music, while Parry used his time writing up his journal, and considering plans for an expedition across the interior of Melville Island when sufficient daylight returned.

Before the sun could be seen again a few cases of scurvy were diagnosed; a Mr Scallon, the gunner on *Hecla*, being the first. Lemon juice as an antiscorbutics was administered and a tray of mustard and cress was sown using soil bought for this purpose, but which only gave a six-day supply. Mr Scallon recovered and Parry wrote, 'I shall be most thankful, should it prove that we can cure, or even check this disease by our own resources: for I cannot help feeling some confidence that, under Providence, our making of the North West passage depends on it.'

The discovery of wild sorrel in large quantities on Melville Island proved to be one good solution. This was boiled and added to the tinned pea soup they had brought with them. The men did not object to the change.

His undisclosed plan, should continuation by sea prove impossible, was headed, 'Plan of a Journey from the North Coast of America towards Fort Chepewyan, (Hudson's Bay Company) should such a measure be found necessary as a *last resource.*'

Their small observatory provided a store for the telescopes and other instruments which had to be regularly checked and calibrated against one another. Sabine prepared astronomical charts using lunar fixes, star sights, movements of comets and the size, shape and position of the Aurora Borealis which appeared mysteriously in different parts of the sky. His charts were published as appendices to Parry's voyage account as were the terrestrial magnetism studies commissioned by the Royal Society (see Appendix 1). Like Parry he also had a keen interest in geology and searched for samples when they could find any holes through the ice. Both identified a selection of minerals including granites, coal, several types of limestone, basalt, quartz, felspar and madrapore.

Beechey sketched the bleak surroundings and many of the animals and birds shot during hunting trips. One of his watercolour sketches, *His Majesty's Ships Hecla and Griper blocked up by snow in Winter Harbour.* (See. Fig. 10) was engraved by William Westall and remains one of the most evocative visual depictions of Arctic 'otherness'. The strength of this subject comes from the stillness, silence and immensity of the star-strewn universe above the ships. A faint representation of the Aurora resembling a delicate mediaeval fresco painting above them with the complete isolation of the two ships resembling large tombs lying under their shrouds of heavy canvas. The engraving has a powerfully religious overtone.

The observatory is visible to the right and an arctic fox, attracted to the camp fire, is the only sign of wild life. The fox is watching a man tending his fire which is thawing out supplies of vinegar for use as an antiseptic. The scene represents the very edge of the known world and has been used in many publications, panoramas and magazines, some with the dramatic title. 'The killing glitter of the stars'.

Lieutenant Edward Hoppner, a talented artist, also made sketches of animal life and both he and Beechey completed a number of beautiful representations of the flora and fauna found on Melville Island. One subject however seems to have been beyond their individual or combined competencies.

The Aurora, as their written descriptions explain, 'was normally visible for upwards of half an hour at sunrise and sunset. It's appearance confined to within a radius across twelve points of the heavens – from SE by E to W by N.' It is 'different and peculiar' on each occasion it appears with strange fluidity in movement as well as in depth and vividness of its blueish purple colours.[5] Sabine attempted to define and explain it on 15 January 1820, writing 'as the atmosphere was clear and serene' we were given a 'brilliant and diversified display, and that the light the Aurora produced could not be compared to that of a moon because it was diffuse and cast no shadow'. He continued by proposing that it 'most resembled the combustion of phosphorus', a helpful piece of scientific observation.

Several crises took place at Winter Harbour during the mid-winter of 1819/20. The observatory burned down and had to be rebuilt under the worst possible conditions. Men doing the work suffered badly from frostbite in their feet due to the thinness of their boots, which then caused mental confusion.[6] Exercising on deck became impossible and more snow made even short excursions ashore too risky, so everyone was confined to the mess decks and cabins.

With only oil lamps and candles to lift the gloom and a large number of bodies concentrated below deck with little circulation of fresh air, problems gradually developed from excessive condensation. Bedding and clothing became damp and unpleasant to wear, as well as difficult to dry out. An even greater problem arose when some of the precious stock of concentrated lemon juice burst their containers in the extreme frost and was destroyed.

By late January the ice around the ships was seven feet thick in a total depth of thirty feet. Isolation, lack of activity and even mild depression began to affect everyone, including Parry. The death of their good-natured ship's dog Carlo upset him badly and he worried about his family at home in Bath. Liddon was unwell which also caused concern and left him without his most senior second officer, with whom he could discuss the situation and his problems. However, he knew he could depend on Sabine, Beechey and Hooper for advice, although none had the same experience of ships and men as himself.

The general mood is best judged from words he wrote:

> not an object was to be seen on which the eye could long rest with pleasure, unless when directed to the spot where the ships lay, and where our little colony was planted. The smoke that there issued from the several fires, affording a certain indication of the presence of man, gave a partial cheerfulness to this part of the prospect; and the sound of voices, which during the cold weather, could be heard at a much greater distance than usual, served now and then to break the silence that reigned round us, a silence far different from that peaceable composure which characterises the landscape of a cultivated country; it was the deathlike stillness of the most dreary desolation, and the total absence of animated existence.[7]

The monotonous and dreary landscape affected their spirits adversely and the routines needed to be given extra momentum to pass the long days. Literacy classes had been started since many of the lower deck sailors had no proper education.

Parry passionately believed that it was his duty to improve the literacy and numeracy, and the personal conduct of all British seaman. Raising educational standards, he maintained, was in the best interests of the Royal Navy and teaching good manners, moral conduct and self discipline was the only way of building a willing and effective crew. Harsh punishments, prevalent on most naval ships were not, he believed, likely to obtain lasting results, only resentment and disaffection. His religious conviction was expressed through demonstrating Christian standards by personal example and teaching from the Bible.[8] Hooper, the purser on *Hecla*, who had been with Parry on the *Alexander* in Spitzbergen, proved to be sympathetic to these principles and became a capable teacher.

When February finally came they had a few more hours of daylight. Small hunting groups were sent out to look for game such as musk oxen, reindeer, geese and small lemmings. These animals had been scarce until then and he admits that the hunting was more a distraction for the men than for finding fresh meat. Coal for the special stoves was running low and the temperature in his cabin over the stern of the ship was only −2°C – below freezing! The lowest temperature taken by Hooper was on the 1 March 1820, and was −29°C at midnight, −11°C at midday, and a daily mean of −19°C.

The amount of fresh snow falling in April was more than had fallen during the rest of the winter and it soon buried the observatory with Sabine trapped inside. After a miserable night alone he was dug out none the worse for his experience reminding everyone of the insulation provided by thick snow which had helped save his life. The sun reluctantly returned above the horizon bringing along with it peculiar atmospheric phenomena created by its more powerful rays: parhelia or 'mock' suns, encircled by several huge arcs of prismatic light beautifully coloured like rainbows. One extraordinarily complex phenomenon was sketched and described later as, 'Parry's Arc'.[9] (see Appendix 2. Meteorology.)

Even by late April the ice remained thick and immoveable. Parry began to worry they might have too little time left to refloat the ships and complete the passage; by the middle of May temperatures had only risen slightly above freezing point during the day and he decided he would prepare the ships ready for sea to keep the men busy and optimistic, and to be ready to leave at the shortest notice. The work was left to Liddon and Beechey while he took eleven of the fittest men to visit the central and northern parts of Melville Island and survey its western coastline. His preference for physical action partly stimulated this as well as his desire to 'see for himself'. The expedition involved hauling with them 800lbs of rations, tents, water, food and fuel, as well as surveying instruments. Six men were harnessed up like horses to pull a wooden handcart made for this purpose and backpacking their bedding and food. Scoresby's advice was that sledges were the best conveyance over ice.

They set out on 25 May, crossing a snow-covered and featureless plain, and reached the far side of the island in under a week. In some sheltered places they gathered wild sorrel, saxifrages, numerous types of grasses and the blue poppy. (See Appendix 2. Botany.) The range and quantity of the vegetation surprised everyone, appearing so soon after months of thick snow and ice.

Snow blindness from the dazzling reflection off the snow badly affected several of them and they soon realised that travel by night was the best answer. Rests were taken under blankets stretched over a rope suspended between two boarding pikes and their cart providing a windbreak. During the journey to the south west they managed to shoot grouse, ptarmigan and a duck. A gap between two hills appeared where they rested and Parry wrote:

> we hastened forward to the point of the nearest hill from whence the prospect was extremely grand and picturesque.
>
> We were looking down from a height of eight or nine hundred feet, onto an extensive field of ice, from which out to the west, we could see no completion for a distance of five or six leagues, about 35 kilometres or 25 miles. The prospect to the eastward being obstructed by more hills. The latitude here was 75 deg 12' and 50" the longitude was 125 deg. 12' and 22".

Originally they assumed this area of frozen water to be a lake, but the shore was the edge of the Beaufort Sea and in the far distance, barely visible, was the eastern end of Banks Land, named in honour of the President of the Royal Society.

The frozen Beaufort Sea was part of the Northwest Passage completed in 1854 by Captain Robert M. Le McClure of HMS *Investigator*. He sledged across from Banks Land where his ship had been trapped and discovered the cairn left by Parry and immediately knew they had completed the same route Parry pioneered in 1820.

Hecla's sailors returned to Winter Harbour without the cart, which had fallen apart, and reached the ships on 5 June. By this time it was hoped that the spring tides on 11 July would free them, but these came and went without the desired result. Parry writes, 'it is not a pleasing truth that we have now been in this harbour, frozen up, during a part of every month in the year except one [August]'.

This new delay was a serious upset to his plans and he now realised could not be hidden from the crew. Would they ever be released from the ice? Could the reduced stocks of food be stretched out long enough to get them home? Would many more men become sick and, if trapped through another winter, what fuel for the stoves would they use? His officers and Sabine felt that the months, even weeks now left for completing their commission would be insufficient. Even if they managed to free themselves during June the chances of getting through the concentration of huge ice floes and 40ft high pressure ridges they had seen in the Beaufort Sea was low.

By turning south they might find clear water, but this would take longer and the chance of finding fuel and game would be more uncertain than where they were at present.

On 24 June the ice around the ships started to break up and move offshore. After days of effort the crews freed both ships by sawing round them, but perversely the sea ice came in once again. During the final few days of June and most of July they continued to prepare the ships for sea and on 24 July Parry ordered the sails to be bent onto the yards. His account giving reasons were made in a stark manner:

> The truth is that it becomes almost necessary to make some show of moving, however unpromising the ice looks, – for it is impossible to conceal from the men the fact that in 8 or 9 weeks from this time, our season of operations in these regions must once more come to a conclusion and it is better to make it appear as long and as bustling as possible.

Parry wrote in his journal that after 14 September exploration at this latitude became rapidly more difficult.

He now kept his men extremely busy and everything was ready, when, finally they were able to leave Winter Harbour on 4 August.

Sailing with great difficulty in thick fog they felt their way cautiously as far west as Cape Providence on the westernmost point of Melville Island, close to the position they reached the previous summer. Parry, Sabine and Edwards walked for seven miles along the shore but quickly realised that further progress was impossible. Ridges of pressure ice twenty feet high stretched across the Beaufort Sea. It was impenetrable and unlikely to dissipate for weeks, even if the summer temperatures were abnormally high. Hooper, as purser, was given the task of estimating how long their stocks might last since there were only four weeks left before the start of the next winter.

Food, if they all went onto half rations, would last one winter, but coal was in short supply and to survive both crews would have to live on the same ship bringing health risks from overcrowding and lack of ventilation.

On 23 August, after discussion with all his officers, Parry decided to return down Lancaster Sound, cross Baffin Bay and set course for Britain. Despite the disappointment everyone felt his judgement was that neither ship could get across the Beaufort Sea in its current state. A more southerly route must be found. Even with reduced rations and conserving fuel stocks they were insufficient for another very long winter trapped by the ice and he came to realise how fortunate they had been to break out from Winter Harbour at all. The welfare of his crew was uppermost in his mind, especially as Liddon and other men like William Scott, were suffering from gastric problems. Scurvy too might break out as their fresh meat and antiscorbutics were finished. Realising he would be asked about any decisions taken to turn back, he knew it was time to return to England, re-crossing Baffins Bay the same way they had come.

By a remarkable change in their fortunes the surface ice vanished, the wind turned in their favour and they reached the entrance of Lancaster Sound in six days compared to the outward journey of five weeks. The advantage they had from such a fast passage through Lancaster Sound was put to good use by staying to correct charts prepared by Ross in 1818 on the western side of Baffin Bay. Parry felt that these had been 'wretchedly manufactured'. They went back into Cumberland Sound, landed on the Bylot peninsula which they realised was an island, and surveyed this in entirety according to their instructions. The severe hardships they had undergone over the previous ten months did not discourage them in in their work, or the knowledge of how much they had already achieved. They deliberately left little opportunity for the Admiralty to say they had turned away and run for home before it was absolutely necessary.

A meeting with three whalers provided an opportunity to send back despatches to John Barrow and Croker, and a letter to his parents, hoping these would all reach England before them. The *Hecla* and *Griper* then continued mapping Baffin Island and visiting Clyde Inlet where an observatory was set up to record an eclipse of the sun. This place was the home of Inuit people and had been mentioned by 'those dear old fellows Baffin and Davis'. In mid-October they finally set sail for England with all his original crew bar two men who had died near to the end of the long winter, one from scurvy and William Scott from pneumonia,

The passage back across the Atlantic was stormy and wild. They lost topmasts and a bowsprit before reaching Peterhead in the first week of November. His exuberance at being back home is shown in the letter sent to his parents with, 'thanks to God for their successes', and showing his absolute certainty that 'the North West Passage is more alive than ever!'[10]

Parry regularly acknowledged the exceptional bravery, endurance, and positive attitude of his officers and crews on this voyage. They had survived some of the most threatening conditions found on earth, including ice in many combinations and of exceptional strength, storm-force winds, thick fog and extreme cold. All in wild, hostile and uncharted seas. They had, remarkably, survived the mental dangers from their ten-month imprisonment in semi-darkness in Winter Harbour without any certainty of returning home. No sailor had been reduced to an unfit state not even by the common cold except for a few minor cases of scurvy. Parry explained to his superiors how his men's resourcefulness, stamina, good health and humour had contributed to the success of overwintering in high latitudes and that there had been no cause for physical punishments of any kind.

For the British press, among friends and fellow officers, he referred to his sailors in his typically understated way as, 'no common men'.

Chapter 7

Fame and Family

The safe return of the expedition received high levels of interest and attention from the public and the press. Despite not finishing the entire Northwest Passage, as Parry said, they had 'knocked a large hole in it.' His promotion to captain was quickly confirmed by a letter that 'expressed satisfaction at his return and at the extensive addition which this voyage has made to the knowledge of the Arctic regions'. For the young officer he still was, his successful voyage gave him recognition as first explorer in a new polar age. Along with Franklin both became popular heroes and the lasting proof of British maritime supremacy, courage and pursuit of scientific excellence.

Arctic and Alpine landscapes proposed both real and imaginary terrors and mysteries, and were a central part of a late eighteenth- and early nineteenth-century fascination with untamed nature. They were integral with the natural sublime, asymmetry that violated all classical canons of artistic regularity. 'Worlds that puzzled, amazed, astounded, and enthralled by their differences to the usual world.' This sublimity was chosen by philosophers for different visions about creation and to develop an awareness for the Arctic being 'the darkest and most imperfect parts of our map'.[1] The public were intrigued by famous authors mostly ignorant about the Arctic including Lord Byron, Samuel Taylor Coleridge and Mary Shelley, and works that fused purity, spirituality and sublimity as one and imbuing at times this combination with mystery, fear, danger and deep personal significance.[2]

Set in scenes of awful desolation, icebergs became an Arctic meme. Their tremendous size and hidden depths, radiant colours and sculpted shapes were evidence of greater powers, and an expectation of sudden and inevitable death. The region, as Byron had commented, 'all expands the spirit, yet appals'.

Over the empty landscape loomed the universe clearer and sharper in cold northern conditions than anywhere else on earth. Shifting mist and cloud moving across frozen seas created atmospheric tricks with the sunlight and moonlight. The unearthly flickering movements of the Aurora Borealis were also part of this imaginary stage on which heroes and monsters paraded: Childe Harold in *Childe Harold's Pilgrimage*, the *Ancient Mariner*, and the powerful fictional creation of Mary Shelley: *Frankenstein. The Modern Prometheus* published in

1818. At the end of her novel its hero, Captain Walton, is conscious of the real nature of the arctic sublime. A hemisphere in which men can 'expand and exalt up to a point but finally what is human and finite in us, terrified on the brink of the mysterious abyss of nature, will draw back trembling from the inhuman and the infinite'.[3] Besides verse and popular novels, the new panoramas shown in cities throughout Britain during the first half of the nineteenth century allowed the public to gaze and wonder at theatrical reinventions of glaciers, crevasses, icebergs, ice islands, and avalanches. Newspapers and journals extolled the idea that the new explorers were befitting of a great nation and that Britain would now stand even higher in the estimation of the world for her determination and enterprise.

Discoveries by Parry had been made in several scientific fields, and Barrow and the Society were certain they could now persuade Parliament, with Melville's support, to vote for new funds for more scientific research and the early completion of the passage. The Russian threat was raised by Parry once more and with one potential route pioneered, the Croker Mountains discounted, and the British whaling industry benefiting from the opening up of Lancaster Sound, the chances for being granted new money were very promising. They had reached 113 deg. 46' west at Cape Providence and the stage was now set for final success. John Ross was quick to write to Parry offering his congratulations but Parry, suspecting his motives, did not respond since he supposed there was a 'game he wished to play'.

The £5,000 parliamentary bounty for reaching 110 degrees of longitude which Parry and his men had won was now added to by an offer of £1,000 (£90,000) for sale of rights for his voyage account to the publisher John Murray. This extra money was very welcome since, even with eighteen months' back pay, he was badly situated financially and expected months of wrangling over his bounty before it was paid out. Initially he thought he would take his time completing his account, but, hearing that Alexander Fisher, the assistant surgeon on *Hecla*, had begun writing his own, he decided to proceed at once.[4] This led to one unhappy consequence, as the collation and analysis of the scientific data could not be rushed and would thus have to be published separately. This decision created an unfortunate rupture in his friendship with Edward Sabine which, although relatively short lived was initially increased when Sabine's niece, Miss Browne, accused Parry of jilting her.[5]

Parry was a national celebrity and his time was occupied by invitations to dine among the highest levels of society, attending as the guest of honour special functions celebrating his achievements, speaking engagements and a command from the new king, George IV, to attend a court presentation or levee. Newspapers and journals continually asked for his portrait to accompany articles about the voyage. As a man who did not welcome all this notoriety, he had to agree eventually to sit for two artists, as well as Tom Garnier – a friend to whom he paid 20 guineas for an engraving for the *European Magazine*. The artists were Samuel Drummond RA, commissioned by Joseph Martineau, his brother-in-law, and the other George Richmond RA for 10 guineas, which he paid himself. (see Cover illustration)

Writing to his parents, he believes that Drummond's was a far better likeness than the portrait commissioned by Miss Burney from Sir William Beechey in 1818, (see Fig. 5) which cost ten times as much.[6] Richmond's portrait shows Parry leaning on a capstan draped with the naval ensign. His sextant rests alongside him underlining his skill as a navigator, and he is holding his telescope, while on his left hand there is a special sealskin glove used in the Arctic by whaling men. In the background an iceberg is moving past on a turbulent sea. He looks younger than his thirty years, with his thick chestnut hair cut short and a boyish face showing almost no signs of strain from his recent expedition. The eyes were an unusual grey, while his mouth and facial expression show evidence of his determination, confidence and natural authority. Still unmarried, as Sir Clements Markham later observed, he represented 'the beau ideal of an Arctic Officer', and was much admired by women.[7]

Letters to his family during these hectic months in London contain more references to what he honestly felt about his success. Phrases such as, 'favourable circumstances played a significant part during the expedition', and, 'the blessings I have received on this voyage are such as deserves the highest gratitude to that Power in whose world we live and die'. His deepest conviction was that Divine Providence had ensured their survival and this conviction was also expressed in his private letters and journals. Many of his explorer contemporaries also felt some powerful 'deity' or 'providence', watching over them.[8] Good luck had been important in his success, which he recognised, but it was largely his meticulous preparations, crew selection, and calm resolve when faced with dangers or unexpected problems, that had made the difference.

He took a decision, with Maxwell's prompting, to write to Melville using a sentence intended, it might seem, for Parliaments attention that 'nothing short of a result by completing the Passage will now satisfy the public'.[9] He proposed another expedition sent this time on a more southerly route, to the western side of Hudson's Bay using Cumberland Sound as a starting point, since the Lancaster Sound passage was too far north to be navigable throughout the entire year. His proposal he admitted meant, 'we have thus to begin again', but argued they should all be sanguine about their chances because, 'the existence of this [the passage] is nearly as uncertain as it was two hundred years ago'. To instil urgency on the Admiralty's part he asked Maxwell to draw their

Lordships' attention to new reports circulating in the Colonial Office, about a Russian initiative intending to claim Russian 'exclusivity of use' in the Arctic Sea bordering the Siberian and the Alaskan Coast.[10]

Parry and Maxwell had dinner again in Pimlico to discuss potential officers for any commission given him the following year. Maxwell had several suggestions, both for extra material resources and Parry's authority over appointments which he needed to insist upon.

Behind the scenes Maxwell worked hard to get Matthew Liddon's promotion, as well as the midshipmen Francis Crozier, James Clark Ross, and William Hooper, his loyal purser. This Parry knew to be very timely since Hooper was helping him with his book.'

The preparation of his account became a trial, and could not be finished until March at the earliest. Working hard at home he was interrupted 'by the distraction of the door knocker going from morn until night!'[11] Despite constant demands on his time Parry managed one visit to Bath for Christmas day, followed by a brief stay with his sister Maria, married to Reverend Thomas Garnier, at Bishopstoke Rectory near Winchester. In a letter sent back later he is quick to explain that he rushed away because he was, 'dining at the Alfred Gentlemen's Club today, the Royal Society tomorrow, and at the Traveller's Club the day after.[12] The first and last, he writes, are considered the 'first Society in London'. His life was bound up with London society which he seemed to enjoy considerably and making many connections over those hectic months which he knew must help him later when any promotion to higher rank was sought. Promotions that could provide better financial security which had became increasingly important not just for himself, but also for his close family.

With his father incapacitated by his stroke and gastric complications, his mother and three unmarried sisters, Caroline, Gertrude and Matilda, all required financial support. Charles, his elder brother, had his own large family to keep and was struggling with the farm and house at Summerhill which were expensive to maintain. Several property investments by Caleb had been unwise and had to be disposed of to pay debts. Parry wanted to support them even though he had a compelling desire to get married – without, at the time, any particular woman in mind. Being on naval half pay he had only a reduced income and no savings. He still worried about the six children of his late sister Emma Elizabeth. Parry found Eardley-Wilmot still undecided over the care of Emma's children, as well as uncommunicative, and wished that his younger sister Caroline could take charge of them rather than some unknown hired nurse, but Eardley-Wilmot still would not agree to this.

On 30 December 1820, soon after his stay in Bath and only two months after his return to England, Parry was commissioned to lead the next expedition in

HMS *Fury*, another naval bomb vessel with a crew of sixty, and his old ship, HMS *Hecla* to be commanded by Lieutenant George Francis Lyon with a crew of forty-seven.

George Francis Lyon had no previous experience of the Arctic, and was the protégé of Lord Bathurst and John Barrow. A moustachioed young man of 22, of medium height, with eyes of an unusual grey like Parry, heavy eyelids and good teeth, Lyon also had a talent for difficult languages. The Admiralty had recently sent him on an expedition over the Sahara desert hoping to reach the Congo river and to travel upstream to find the source of the Nile. Lyon was the only surviving officer and Barrow agreed with his new appointment as an easy way of keeping him silent about the Congolese failure. Lyon was an extrovert, irrepressible in spirit, and a fine draughtsman. He held anti-slavery views and believed that all Arab people should be educated to dissuade them from future slaving activities.[13] The interests and skills both he and Parry shared were instrumental in the strong relationship that developed despite their different personalities. William Hooper, the efficient purser, had been promoted to lieutenant, James Clark Ross and John Edwards were also going on the *Fury*, while Henry Hoppner and Reverend Fisher, chaplain and astronomer, went with Lyon on *Hecla*.

The ships were made ready to leave in April but before departure Parry was presented with the Freedom of the City of Bath on the 24 March.[14]

He received his citation in a wooden box made out of panelled oak from *Hecla* which read: 'presented to William Edward Parry Esq ... as a token of their high sense of importance of the late discoveries in the polar regions and the many eminent qualities displayed by him during that perilous expedition.' Exactly one year later while he was away, his father Caleb died of a heart attack.

In Caleb's obituary, the *Bath Weekly Chronicle and Weekly Gazette*, referring to William Edward Parry as 'his beloved son preserved during a voyage of unparalleled enterprise, danger and interest and of whom he, Caleb, providentially could witness the national applause, the universal admiration and well-earned honours that were his rewards.'[15]

The voyage account was printed in quarto volumes with engravings by William Westall, and was published in May 1821, just as his new expedition was moving slowly down the Thames to the Nore. It sold out rapidly with extracts reprinted in the *Gentlemen's Magazine* and elsewhere. The journal with its comprehensive scientific contents built on his growing reputation in both Europe and Britain. The evocative illustrations were widely admired for interest, objectivity and clarity, all being carefully worked interpretations of coloured sketches by Frederick Beechey and Edward Hoppner.[16] In his piece for the *Quarterly Review*, John Barrow made several flattering comments about the geographical discoveries, the well chosen and executed illustrations, and most of all Parry's lucid narrative. Parry wanted a special commendation for Sabine, 'Captain Sabine supplied the place of a person of the description, professional naturalist, and to him *in particular* as the Appendix will show that science and philosophy stand greatly indebted for a collection of facts and experiments in a part of the world hitherto but little known and never before visited by Europeans.'[17]

Barrow's positive review was also directed first and foremost against John Ross: 'In this work we find *no* display of self-importance, *no* attempt to deceive, or throw dust in the eyes of the public; *no* marvellous stories to disgust or confound and make the ignorant stare. *No* representation of objects the mere fancies of the brain.'[18]

At Deptford Dockyard Parry and Lyon worked on their ships, making them stronger for the ice conditions they anticipated. The hulls were doubled giving more strength, and larger and stronger rudders with modifications for raising and lowering them when required were newly installed and tested. The ships were rigged identically since having the same draught allowed the interchangeability of spars, cordage and other gear if damage occurred to either one. Innovative insulation techniques had been worked out, including special cork bulkhead linings for the mess decks to help reduce condensation; a new type of stove was placed lower down in the hulls that ducted warm air throughout both ships transmitting warm air upwards. The galleys were turned about to face aft, giving the crew's quarters extra warmth as well as a more cheerful appearance, and a special ingenious snow-melting vessel, designed by Messrs Lambe and Nicholson, was attached around the galley chimneys. This device supplied sixty five gallons of pure water from snow fed in from above and drawn off by a tap down below, even with temperatures well below zero outside. Clothing included tough canvas shoes lined with wool and soled with hide, deerskin jackets and several experimental types of snow goggles were adopted. Parry wrote later that nearly all his ideas worked well, although they had never been tried before.[19]

Salt pork was substituted for beef and the newly invented corned beef was tried. Supplies of rum, lemon juice and vinegar rations were concentrated, saving space and being much less likely to deteriorate in storage. Canned foods, including soups and vegetables, were prepared in quantities sufficient to last three years.

Special kiln-dried flour, sealed in airtight wooden casks, was loaded as well as the nearly indestructible ship's biscuit! To save weight on the stormy Atlantic crossing an extra quantity of coal was transferred to another ship to be taken aboard when they reached the entrance to Hudson's Bay.

Many of the men who had been with Parry previously applied again. There was no shortage of eager crew members. Lieutenant James Clark Ross was appointed as the replacement for Captain (now Colonel) Sabine and at the last moment, the Royal Society decided to send Reverend George Fisher, a surveyor and a chaplain, to help with the pendulum and other scientific experiments. He was the result of the Admiralty Board's cost-consciousness at all times in refusing to help remunerate Sabine themselves. Parry asked for, and got, the experienced George Reid as his ice master.

Hooper describes the crews as 'stout able bodied seamen', and was sure that Parry's consideration for their welfare gave them all exceptional confidence, despite knowing the terrible hazards that lay in the future. He noticed how Parry addressed each man by name and considered them as one family, each dependent on the others but all individuals with various needs.

The distractions and amusements tried out previously were prepared once again and Hooper describes how, 'trunks of theatrical costumes, a printing press, a new epidioscope for lantern shows, a full library of books and Bibles for everyone, were all loaded aboard'. Over a hundred scientific and navigational instruments went with *Fury* and another forty-seven on *Hecla*, including thirteen different chronometers, twenty-four types of thermometer, many hydrostatic balances, and the latest type of astronomical clock patented by a Mr Barrett which was cared for by Fisher, the Royal Society representative.

George Lyon and Charles Bushman, a young midshipman, were to make sketches and drawings of all subjects of 'unusual interest', as well as clear and comprehensive records of harbours, inlets and landmarks of significance.

The official instructions were unequivocal.[20] They were to enter Hudson's Bay through his Strait until they reached Repulse Bay, or a part north of the Wager River, which they decided belonged to the continent of North America. Then to sail north along that coast surveying every bend and inlet. At the end of this exercise to continue exploration to the west and north, and at the end of the summer decide whether to remain longer or return to England. 'Your experience and judgement of the past induce us to *trust* this point to your own discretion.'[21] The instructions ordered to cultivate relationships with any natives they met, and, if they were able to obtain a position far enough up the northern coast of America (Canada) they were to erect a large flagstaff in a prominent place. Underneath this they were to seal up and bury detailed information on the route taken and any advice for John Franklin and his men on finding their way back to Hudson's Bay and the Company forts. They were to take his group on board as 'supernumeraries' if they joined up. As these orders were being given to Parry, Franklin's party were waiting for the ice to thaw at Fort Enterprise, 250 miles south of the Coronation Gulf on the Polar Sea, and eighteen months before they finally managed to return to England.

Although the discovery of the passage was still the main objective at the Admiralty they responded to Royal Society interests by adding to their instructions, 'your observations etc. must prove most interesting and valuable to Science. We therefore desire you to give your *unremitting* attention and to that of all officers under your command to these points as being objects of *the highest importance.*' George Fisher was presumably chosen to see that they did!

The *Fury* and *Hecla* put to sea on 8 May 1821 having welcomed many visitors as well as holding a Grand Ball called a 'spree', hosted by Parry, to celebrate their departure. By the time this was over he was exhausted with the crowds, the questions and frequent requests for interviews. He wrote in his last letter home, sent from Greenland a few weeks later, that, 'I long to be in the ice', and admitted in a reflective and pessimistic mood that, 'I never felt so strongly the vanity, uncertainty and comparative unimportance of everything this world can give and the paramount necessity of preparation for another and better life than this.'[22]

Chapter 8

Illigliuk and Igloolik*

The North Atlantic crossing took eight weeks from the Orkney Islands with fierce gales forcing them to turn back twice. The Greenland pack ice was very heavy and slowed progress so it was 3 July before they reached Resolution Island at the entrance to Hudson Strait, far later than Parry had planned. During these weeks he watched the performance of *Hecla* and *Fury* and was fully satisfied they were better for cruising and surveying together, whereas *Hecla* and *Griper* were not, the latter usually struggling to keep up with the former. Surveying depended on regular communication between both ships and Parry now felt confident that the vast area they had to cover might yet be completed during the few remaining weeks of the summer. At Resolution, *Ay-wee-lik* in Inuit dialect, they transferred extra supplies from the *Nautilus*, a cargo ship sent along to prevent overloading the expedition ships across the Atlantic. These comprised more coal, tinned foods and twenty bullocks.

In 1742 Christopher Middleton had explored the western part of Hudson's Bay but was unable to pass through the icebound and narrow straits at an opening to a large bay he named Repulse. The northern side of this had scarcely been visited and almost no hydrographic information existed. This posed the inevitable question as to whether there was a navigable gateway beyond Repulse Bay into a north-west passage. Parry's orders were to investigate and map this coastline, exploring every inlet discovered.

The seamanship of George Lyon, his second in command, had originally been of concern as his recent exploring work had been in the deserts of North Africa, but Parry quickly realised Lyon was capable of ship handling in poor weather, had determination equal to his own and the physical stamina to operate small boats even under the worst possible conditions. Parry soon developed a confidence that Lyon could work independently and had no desire to 'tether' him, as John Ross had previously done to him. More territory could be covered between his two ships as a consequence.

Lieutenant Hooper, his purser and friend, remained on the *Fury* together with Reverend George Fisher. They both held strong Christian beliefs and shared

---

* A rough translation of the Inuit dialect is 'There is a house here'.

the same view that importance should be attached to the teaching of good morals and better general education for ordinary seaman. Fisher's knowledge of astronomy was not as complete as Sabine's but he was enthusiastic and diligent, despite suffering badly with seasickness throughout the voyage. Hooper's chief interest was ornithology and so kept lists of birds which he compiled with precise descriptions and sketches. James Clark Ross did likewise for any mammals and small insects and performed taxidermy on those they wished to keep. Records were kept of the body temperatures of each victim.

The crews were repeatedly drilled with all the sails to improve the speed of changes needed in changeable weather or in avoiding sections of ice. The men were crammed into smaller spaces on their mess decks because of the extra stores they carried and the officers watched for any sign from the newer recruits that they might mentally or physically deteriorate during the long winter. They would live with less than two hours daylight during January, sub-zero temperatures day and night, and under close living conditions.

During the Atlantic crossing Parry had set a course first taking advice from the captain of a whaling ship. He expected the winter ice would form during October, as it had done two years ago in Lancaster Sound.

Speed was now essential and he took the direct route, staying north in Hudson's Straits where there was less tide, and then passing through a dangerous area strewn with rocks and shoals among the Savage Islands. From there they would steer north-west to Southampton Island and into the Foxe Channel looking for Middleton's 'frozen straits' and the entrance to Repulse Bay.

On reaching the Savage Islands they were met by a number of Inuit in large canoes they called *Oomiaks*, which when translated means 'luggage boats'. These were 20ft long and used for transporting people between their summer and winter settlements. They were full of women. (Fig. 13. Canoe of the Savage Islands.) Lyon calculated there were nearly one hundred people in total offering goods for trade: tusks from walrus and narwhals, skins from seals and deer, gloves, and harpoons with ivory points. In exchange they wanted nails, beads and knives. The women's clothes were well ornamented and the people were 'of good stature, well-proportioned with small slender heads and feet, and broad visages'.[1] But Parry also noticed that they were very dirty and showed an alarming propensity to bleeding from the nose. They were, he said, 'marvellously given to thieving especially of iron'. When a transaction was agreed the Inuit concerned would lick any object first to signify that it was now theirs.

After the minor islands they sailed north for Salisbury and Nottingham Islands and then north-west, avoiding some large icebergs, and reached Southampton Island on the North Eastern coast on the 4 July in swiftly changing weather conditions; heavy snowstorms followed by belts of fog and then, abruptly, unexpected periods of bright sunshine. Here they went ashore and collected rock samples; gneiss, epidote, feldspar, quartz, garnets, graphite and sillacks.

Hooper wrote his opinion that these were all identical to those found at the Prince Regent Inlet. Lyon tried to identify and dry various plant specimens including a yellow poppy, white andromeda, an unusual buttercup, sorrels and a dwarf willow that grew to only 3ft in extent, keeping itself quite flat to the ground.

They believed they were now at the entrance to Repulse Bay, but in this they were mistaken. They were inside a fine natural harbour, landlocked on three sides. They named this after the Duke of York, before moving to Southampton's northern end in mercifully clear weather. It was an easy passage and they came through the frozen straits without incident. They had now entered Repulse Bay with no signs of pack ice on any part of the horizon.

Eleven hundred miles over to the west, John Franklin with his small group of British seamen, Dr John Richardson from the Royal Society and a dozen voyageurs, half-breed Indians and white settlers disliked and mistrusted by the indigenous Indian tribes, were paddling and carrying their boats down the Coppermine River following Samuel Hearne's previous route to the Arctic Ocean.

The good weather that had assisted Parry provided time enough to complete a full survey of the western and northern sections of Repulse Bay; an area of over sixty square miles. It was discovered that powerful tides were coming into Repulse from the east, but nothing to match these coming from the west.. This suggested to both Parry and Lyon that Repulse Bay had no channel out into a north-west passage, but they surveyed and charted every small channel to confirm this opinion and continued the work up until the end of September. Parry led several expeditions personally to be certain of his decision before finally leaving Repulse Bay to set a course north-eastwards along the coast of the newly named Melville Peninsula.

Survey work was always very demanding. It involved hauling and rowing their boats through shoals, hidden reefs and rocks. When on shore the surveyors were mostly trekking miles over desolate country and just one especially long and complicated inlet, named after Lyon, took them an entire week to survey.

Parry took a group of nine men to explore the area between this Lyon inlet and the northernmost part of Repulse Bay. They were away for more than two weeks and had still not returned so Lyon, concerned for their safety and whereabouts, took *Hecla* south once more and found the group with their small boats trapped behind fresh pack ice. Only the frequent firing of guns on *Hecla* enabled the stranded group to find them. Two more days and they would have perished.

After six weeks of unremitting effort, and with the thermometers all showing midday temperatures barely above freezing, new surface ice began to reduce daily movements. The anxiety and frustration this created was imaginatively described by Parry: 'A ship in this helpless state her sails in vain expanded to a favourable breeze her ordinary resources failing ... has often reminded me of Gulliver tied down by the feeble hands of the Lilliputians.' Henry Hoppner produced a sketch of them in this situation which was included in the voyage account for the third Expedition of 1824–1826 on page 33. Rain when it fell now froze instantly, making decks and ropes resemble glass. The time to look for winter quarters had come.

At the entrance to the Lyon Inlet was a small island protected from northerly gales and a good position away from shifting icebergs. This was called *Neyuning Kitua* by the local Inuit, Winter Island by the expedition, and became their base for the following eight months. Parry felt the progress made this far was satisfactory. 'I derived the most sincere satisfaction from a conviction of having left no part of the coast from Repulse Bay eastward in a state of doubt as to its connection with the Continent [of America].' The entire complement of both ships began sawing out another canal through the ice half a mile long. For this winter they were 9 deg. further south than at Melville Island and had ice which was lighter. Several men fell through and had to be rescued before freezing to death.

After the canal was completed the crews were issued with 'carpet boots with cork soles, and comforters'.[2]

By now it was 8 October and thick snow banks were built round the ships as protection from the wind and to give extra insulation. Their portable observatory went ashore as well as a separate building called a 'boiling station', in which they could render seal blubber into oil suitable for their lamps.

Another theatre committee was set up and rehearsals began for the opening play of the season which was the *The Rivals* by Thomas Sheridan. Hanging spaces for drying clothes and the airing of bedding were set up around the new Sylvester stoves, and after this work the conditions were easier than those experienced at Melville Island in 1819–20. Lyon, who had not been part of that expedition, wrote in his private journal, 'every arrangement which could contribute to our general comfort and health was made by Captain Parry'. Being further south this winter there would be two more hours of daylight even in January, a helpful respite from the persistent winter gloom. But the monotony of the landscape and the dreariness of the ice everywhere still produced a peculiarly melancholic mood.

Suddenly, and to their great good fortune, sixty Inuit arrived. These were people of the Nunaavik and Nunavut tribes who came from some distance away. They were cheerful and kindly disposed to the visitors and had an openness and honesty which contrasted favourably with the people of the Savage Islands. Their first visit started a relationship that became increasingly friendly, convivial yet respectful. It provided Lyon with many opportunities to use his sketchbook to explain how they constructed snow houses or *Iglu*, dog-drawn sledges, kayaks and umiaks, a number of household items, clothes and their fishing and hunting weapons. On occasion they also allowed portraits of themselves to be made.

This record was one unexpected but valuable consequence from the voyage and as accurate depictions they produced the first proper record about the lives and living conditions of this extraordinary race of people. Despite often being close to starvation they were uncomplaining and willing to help the Kabloona, the visitors, in every possible way.[3]

Lyon's work was exact in every small domestic detail, and shows his admiration for several members of the group. The people were not patronised nor caricatured like the 'Arctic Highlanders', so named by Ross, but shown as personable and interesting individuals. One, a small boy called *Nak-kahlioo*, who took the nickname 'kettle' from the sailors, was very popular, spending nights on *Hecla* acting as an interpreter. Another was a good looking, intelligent young woman called *Illigliuk*, who managed from her local knowledge to construct a map of the island and the coastline going north. (See Map 3. Winter Island drawn by Illigliuk.) Her efforts seemed to confirm that they were at the tip of North Eastern America. Because of her help and advice Parry befriended her family and offered to take *Toolooak*, the son of *Illigliuk* and her husband *Okolook* back to England, but very sensibly they refused.

Her drawing showed a channel fifty miles north of Winter Island that stretched westwards, out further than the Wager River and into a great expanse of sea that *Illigliuk* said could be seen clearly in good weather. This news, which Lyon mentioned in his journal, 'set us all castle building and we already fancied the worst part of our journey over'.[4] Parry was less sure, but he too showed optimism stating that, 'our being very near the north eastern boundary of America that the early part of the next season would find us employing our best efforts pushing along its northern shores'.[5] This renewal of hope raised everyone's spirits during the darkest months of that year and with all his efforts to occupy the men, they were kept busy, cheerful and mainly in sound health.

The sun had dropped below the horizon on 2 December and did not reappear for forty-two days. A new play was performed every fortnight, written and performed by the officers. *Raising the Wind* and *The Mock Doctor* were farces and Parry took parts in both. Their magic lantern was used for *Phantasmagoria* and several other shows, and the officers were invited to musical parties 'in both Captain Lyon's cabin and my own', despite Parry's quarters being the coldest place on the ship. Only the newspaper did not reappear due, it was said, to 'lack of news'. On one occasion and probably at Lyon's suggestion, a concert was attended by five women and one man from the Inuit village.

Mr Halse, *Hecla*'s clerk, started a school for twenty seamen of which, Parry wrote, 'I can safely say that I have seldom experienced feelings of higher gratitude than in this rare and interesting sight.'[6] The pursuit of all these activities induced, 'cheerfulness and good humour as well as sensible occupation,' and in which he believed, 'aids in a supporting manner, the cure for this extraordinary disease' (scurvy). His observation after two winters locked in the ice being that 'passions of the mind in reducing or removing scorbutic symptoms are too well known to need confirmation or to admit doubt'.

Daily tests on the many instruments they carried continued and results compared carefully. They took records of the weather, wind, atmospheric changes such as mist and fog and unusual meteorological events.[7] Parry and Lyon went out prospecting together looking for rock samples and visiting sites that had been used as graveyards by the Inuit.[8] Lyon spent some days living in the Inuit snow houses eating their food and learning enough words to compile a basic vocabulary, eventually becoming so familiar to everyone they called him, 'father of the Kabloona.' The Inuit name for the visitors.

These people could never understand why the crew were unrelated to each other since their own people invariably were, and the idea of punishment for wrongdoing was completely alien; children were neither corrected nor scolded.[9]

Daytime temperatures had now dropped below −35°C and hunting and fishing were impossible, so supplies of fresh food started to run low among the Inuit. Parry provided ship's biscuit and 'pemmican', a nutritious dried meat mixture rich in fat, from both ships. Hooper reported that, 'with our time thus occupied, our comforts so abundant and the prospect to seaward so enlivening it would indeed have been our own fault had we felt anything but enjoyment in our present state and most lively hopes and expectations for the future'.[10]

At the end of March, in expectation that the ice would release them before May, Lyon volunteered to lead a two week expedition following the coastline to the north using their new map. He was restless and wanted to try out his new sledges. To this Parry soon agreed, which may have been a deliberate means of separating him from *Ang-ma-loo-tooing*, a pretty widow, and several other Inuit women.[11]

The expedition was away two weeks and nearly ended in tragedy. Winter was far from over and their clothing and supplies, such as tents and sleeping clothes, were inadequate. They lost their way in heavy snow and several men became disorientated with frostbite and snow blindness and were 'reeling about like drunken men-completely bewildered'.[12] Lyon later confessed to Parry that four of his party probably had less than an hour to live when, miraculously, they saw footprints in the snow that led them back to the ships. Despite this near disaster the sledging experiment was successful, and the records they kept proved that four strong dogs pulling sledges with a maximum weight of 100lbs could travel a mile in six minutes. This was helpful knowledge and of practical value for future expeditions, but it appears to have been discounted or simply ignored by the Admiralty. The use of sledges and dogs learnt from the Inuit was seemingly disregarded on the grounds that these were the practices of 'primitive' people.

When sledges were adopted during searches for Franklin's lost expedition they were driven by sails, flew pennants and were man handled by sailors!

May and June passed but they remained held fast. It was not until the second week of July that they were finally able to break free and start their passage towards the place close to *Amtiok* called *Igloolik*, meaning 'there is a house here' in local dialect. *Amtiok* protected the entrance to a narrow strait leading westwards. On reaching this point they found thick ice reaching from shore to shore and were stopped and prevented from going forward. At this point Parry reached a state of sad resignation. His journal reads: 'we are halted at the very threshold of the Northwest Passage for nearly four weeks. Suspense at such a crisis was scarcely less painful because we knew it to be *inevitable*.' He now organised a small party with sledges to cross the centre of the Melville Peninsula and travel down to the western shoreline beyond the newly named Hecla and Fury Strait. The weather was better and long distances were covered daily. They reached the south western side of the peninsula which was full of deer, fresh water lakes full of fish and was inhabited by more Inuit who were friendly. There were long uninterrupted views from its highest point out to the west where no land was visible. Parry felt sure this was the Polar Sea. 'We were at the point of forcing our way through it along the northern shores of America.'[13]

Returning to the ships with this news there was little else to do but wait and hope that the ice in the straits cleared but the main section remained blocked with ice as solid and continuous as 'mountains 1,000 feet high'.[14] The Inuit had told them that this route was normally clear during summer months, but now everyone was struck with 'disappointment and mortification which could not possibly be described'.

More groups went to survey the far end of the frozen straits and found that the channel passed into an expanse of open water that Parry incorrectly marked as part of the Polar Sea, but was in fact a landlocked sea, eventually named by John Ross as the Boothia Peninsula.

In early September strong gales were coming out of the northwest and, having waited as long as possible, Parry and Lyon agreed to overwinter a second time and hope the channel would be open the following spring. They moved down to a safe place at *Igloolik*, the island they had passed going north.

Igloolik was a seasonal settlement lived in by several hundred people. They too welcomed the visitors, albeit more cautiously at first than those at Winter

Island. Contact was more formal and less familiar, but as Lyon's charm prevailed a regular daily intercourse developed and explanations provided about their daily lives, hunting and fishing practices and unusual customs. Parry wanted to convert them to Christianity, feeling that conversion would bring them benefits and he recognised that many possessed a 'degree of harmony among themselves which is scarcely ever disturbed', and on that basis he judged them to be of a 'naturally charitable disposition'. Their souls, he wrote, given a sound Christian education imparted with Christian principles, might be 'benefited to a significant effect'. His attitude towards these Inuit was completely different to those from the Savage Islands whom he despised as thieves, pilferers and pickpockets, and being 'very foul of habit'. Lyon's view was far more pragmatic, stemming from the time he had spent among nomadic desert Arabs; he appreciated their different values of love, compassion, jealousy, sympathy, gratitude and ownership of possessions. They were, he reasoned with Parry, a direct consequence of life in an extreme climate faced with starvation and death and with almost no resources other than hunted game. People had little regard, he realised, for sentimentality, gratitude, or outward signs of passion, and most of all warlike behaviour.

Unfaithfulness by the married women did not apparently create any anger or resentment from their husbands. Stealing metal implements, which all were given to, annoyed Parry considerably but Lyon reminded him that iron of any kind was their equivalent of a solid gold bar for Europeans. They had walrus tusks, animal bones, occasional pieces of driftwood or deer antlers from which to manufacture their tools but iron being stronger, and when sharpened with stones, was incomparably better for making harpoons and hooks. These people lived from day to day surviving with skills that astounded the expedition, producing structures including windbreaks, light but strong canoes or kayaks, sledges and harnesses, and even a kind of snow goggle. Their code for survival was to help one another depending on immediate needs. 'One day you helped a man and the next he helped you', important in the Arctic where material resources were so limited. Their eating followed this pattern too, so that whenever food was available it was consumed by everyone in large quantities, but nobody complained when it was not.

Food became an interesting but awkward subject between the visitors and their hosts. Items such as coffee, gingerbread, jelly and rum were as much disliked by the Inuit as raw blubber, chunks of seal meat and raw fish were by the sailors. The Inuit diet was a source of fresh vitamins and no scurvy was noticed among any of them.[15] Scurvy on several British expeditions could have been avoided if these rich sources had been prepared and cooked in ways that sailors could consume.

During the next winter theatrical entertainments were reduced, the given reason that these were 'entirely unnecessary' because their Inuit friends, by 'daily visits to the ships throughout the winter, afforded both the officers and the men a fund of constant variety and amusement'. The Bible and other classes continued as usual and just two farces were performed by the officers, *Raising the Wind* and *The Mock Doctor*.

The next two week trek was planned using several dog teams lent by the Inuit, which Parry, in a letter to his family back home in England, praised as 'most invaluable animals in making journeys and in transporting stores from ship to ship'.[16] It was longer than any that had been attempted previously and lessons and experience gained should have been applied during his later attempt to reach the North Pole but never were. Had they been, the outcome might have been very different.

During the second winter, despite the ministering of scorbutics and the diversions derived from the Inuit, some crew members fell sick. Edwards the surgeon on the *Fury*, reported 'a serious decay in the general health of the men'.[17] By August, after eleven months of winter and four months after the anticipated arrival of summer, the ice still showed no sign of breaking up. Tension between some of the officers grew which disturbed Parry and Lyon, but the causes arose not so much from personal enmity, but frustration, boredom, the monotonous food and differences of opinion over the best course of action for them to take.[18]

Determined to carry out all the conditions in his instructions, Parry considered sending only the *Hecla* back to England under Lyon's command, taking the sick and some less motivated officers. He drafted a letter to his parents to be sent with Lyon in which he said: 'I cannot despair of still ultimately effecting our object. I am determined however with the continued assistance of Providence to show that perseverance has *not* been wanting in this enterprise; and no consideration shall induce me to relinquish it while a reasonable hope of success remains.'[19] The criticism levelled at Ross was perhaps uppermost in his mind, and with *Fury* he thought to remain another winter, try to pass through the Hecla and Fury Straits and then reach the Pacific Ocean.

However on advice from Edwards the surgeon, who had now detected several new cases of scurvy as well as the death of five men from causes such as dropsy, inflammation of the bowel and pneumonia, as well as seaman Pringle, who fell out of the rigging, he changed his mind.

On 8 August they left *Igloolik* and after one more look into the entrance of the blocked Hecla and Fury Straits, Parry decided to take both ships back home. If *Fury* had stayed it was most unlikely that she would have been more successful. The crew and the supplies of fresh foods were very low and his final decision, though made reluctantly, was in the best interest of all the men's welfare and safety. The return passage now proved to be more difficult than the outward one and every able-bodied sailor was needed to work the ships.

Initially they became surrounded by moving pack ice which carried them south for 420 miles out through the Foxe Basin and into Hudson's Bay, powerless to prevent themselves being wrecked on shoals or beached on the rocky lee shore. Fog moved around, obscuring these hazards, and winds shifted continuously. It was a serious test of seamanship and navigation.

Twenty-six days from departure, on 17 September 1823, they reached open sea, and by 10 October they were moored off Lerwick in the Shetland islands, where bells were set ringing and the town specially illuminated, 'as if each individual had a brother or son among us'. They had been gone two years and three months.

The voyage account, published in 1824, concluded that:

> the result of the endeavours of the two expeditions lately employed under my orders have at least served the useful purpose of showing where the Passage is *not* to be effected, and of their bringing within very narrow limits the question as to where any future attempts should be made.[20]

This could be read as understatement, as his first expedition had already shown that one direct route was through Lancaster and Melville Sounds, while the second that a southerly and, supposedly, less ice-bound route, did not exist out through the northern end of Hudson's Bay.

The open water they had seen from the western side of the Melville Peninsula would be gained through the Prince Regent Inlet. One route Parry had examined briefly in 1819 but now believed required a further search. 'There is no known opening which seems to present itself so favourably for this purpose as Prince Regent Inlet'.

The British Press were less impressed with this decision. The New Monthly Magazine took a sceptical view. 'The expedition has in fact neither added much to geography nor been able to explore further than was done by Middleton and preceding navigators'. John Barrow was still determined to judge this expedition positively agreeing with Parry, 'there is no known opening which seems to present itself so favourably for this purpose as Prince Regent Inlet.[21]

The recent voyage had produced a reliable chart for northern Hudson's Bay, the Foxe Channel including Southampton Island, the Icy Straits, all parts of Repulse Bay and the treacherous long eastern coast of the Melville Peninsula. It returned with many botanic, ornithological and mineral specimens, as well as proof that the Aurora Borealis did not affect compasses. Due to Lyon's notes, the short vocabulary and twenty-six coloured sketches, there was now a unique and astonishing account of the native *Nunavik* tribes from Labrador and other places in North Western Canada.[22] Barrow was so impressed by both this journal and the beautiful sketches that he proposed that they should be published privately, unusual advice since it gave the Admiralty no further control over its content or use.

Lyon agreed to his suggestion and, compared to Parry's authorised version – by naval convention restrained, factually detailed and unsensational in its descriptions – his work is quite different. His account of the Inuit people and their way of living are devoid of uninformed criticism or any patronisation. The account contains several hundred pages of observations about snow houses, daily activities, family and communal relationships, their treatment of old people, clothing, social customs and hunting and fishing methods. Much of the detail was observed at close hand when he lived in their village. This record and the drawings from the voyage, showed that the Sub Arctic was populated by people, who for hundreds of years, had built up a sustainable existence in the most extreme conditions, and from whom future explorers had a great deal to learn.

Chapter 9

Fury Beach

Parry's return from his second expedition in October 1823 was greeted with more muted national celebration than the first. He had spent two long winters in the ice, surveyed the farthermost shores of Hudson's Bay, and, largely thanks to George Lyon, produced a unique and remarkably well illustrated account of two groups of Inuit living on the Melville Peninsula. The illustrations were as remarkable as any produced by William Hodges during the first voyage of Captain James Cook of Tahitians and South Sea Islanders published fifty years previously. But the appeal of the Inuit people in their ice covered landscape, their physiognomy and very basic lifestyle did not have the same appeal to the British public. John Ross' account of the Greenlander Inuit was also partly to blame. He had carelessly labelled them using descriptions that denigrated their culture and social customs such as pulling noses, valuing simple primitive possessions, eating raw seal meat, fish and walrus, and having only unsophisticated tools. They were presented as people living in the Stone Age! The lack of iron tools did not label them as 'primitive' and his trivialising description, 'Arctic Highlanders', associated them in the minds of English people at this time with the uncouth manners of the Highland Scots. They were neither uncouth, uncultured, nor childlike as he portrayed them, but lived in an evolved civilization with customs and practices using special skills necessary for survival.

Franklin had returned a year earlier to wide acclaim and public popularity. He was, like Parry, harried by the newspapers. They wanted dramatic accounts of the privations he had undergone, threats to their lives from wild Indians and the murderous intentions of their semi-starved 'voyageur' half-breeds, who had been subject to ill treatment and contempt by most of the British officers. The murder of midshipman Hood by a 'voyageur', followed by the execution of that murderer by Dr Richardson, created a thirst for even more sensationalist and lurid events.[1]

Franklin became a national hero before reaching England because the *Times* newspaper had run an article sent from Montreal, their port of embarkation. This announced that his expedition survived chiefly because they 'prolonged a miserable existence by chewing upon the tattered remains of their boots!' Once back in London Franklin was invited everywhere and lionised by the rich and famous; John Murray published an expedition account in June 1823, which became a best seller and quickly translated into four languages. Only two years after first publication, copies were selling for ten guineas (£1,500).

The survival of the expedition was a truly epic tale. The journal prepared by Lieutenant George Back and Dr John Richardson FRS, the naturalist, told an extraordinary story of their gruelling march interspersed with perilous voyages down fast flowing rivers in small canoes through treacherous rapids and finally weaving in among icebergs lurking behind fog banks along the mouth of the Coppermine River. It was a wilderness unknown to Europeans described and illustrated with descriptions of the Inuit tribes, the Iroquois and Copper Indians who mainly acted as guides and hunted game for their meals, and had many illustrations of the features in the landscape discovered en route.

It was a more entertaining read than Parry's account of two seemingly uneventful seasons locked into the winter ice. Franklin contributed little to this journal – not being, he confessed, 'good with such things', and left the work mainly to Richardson and Back.

Authorised by Barrow and Sir Humphrey Davy, the President of the Royal Society, their journal showed the courage and tenacity of Franklin and his small group of British seamen who accomplished this dangerous journey in wild and difficult country.[2] Franklin, a modest, friendly but undemonstrative man had strict Sabbatarian ideals and deeply held views about personal morality. One biographer describes him as 'a curious compound of elevated religious zeal and personal integrity dedicated to the pursuit of naval service, exploration and scientific work as *divinely* ordered duties'.[3] He was becoming deaf because of the cannon fire he had endured during Nelson's great victories of Copenhagen and Trafalgar. The deafness increased his reputation for aloofness, which although unjustified was partly the nature of a shy person. Like Parry, he disliked publicity and excessive displays of patriotic celebration enjoyed by the public. They were private individuals even though Parry enjoyed his adoption into the highest levels of English society.[4]

The two men met several times at Somerset House and Franklin made special visits to London from his home in Lincolnshire for that reason.[5] During the unusually harsh winter of 1823 they were both invited to the same London parties and ceremonies in their joint honour. At one dinner party Franklin's fiancée, Eleanor Anne Porden, introduced Parry to her closest friend, Isabella Stanley, youngest daughter of Baron Stanley of Alderley.

Isabella's mother Maria Joseph, the daughter of John Holroyd the first Earl Sheffield, was co-founder of Girton College Cambridge. She was a friend of Gibbon the historian, and an ardent feminist and reformer. His first meeting with Isabella made a lasting impression because of her 'fragile looks, joyous spirit and acknowledged beauty'.[6] (see Fig. 12.) The first meeting with her mother however was less successful and was reported later by her sisters and brothers as not having gone well.[7] Their differences arose partly from several malicious reports of his previous attachment to Miss Browne, Sabine's niece, and a distant relation of the Stanleys. These had been deliberately circulated by Miss Browne's mother.

Lady Stanley was concerned about his religiosity since she had a contempt for the established Church of England. She did not approve of the Evangelical movement either and was so deeply entrenched in her views that, despite having two ordained sons one of whom was eventually made the Bishop of Exeter, she dismissed both as 'lacking in belief.'

The Parry journal, using some of George Lyon's illustrations, was less successful both critically and commercially than Franklin's, although it was thorough and had a large number of appendices listing new species of animals, mammals and birds as well as the mineralogical, astronomical, zoological and meteorological records kept by Lyon, Fisher, James Clark Ross, Hooper and Parry. The content was so great it delayed publication until after Parry had left on his next voyage. Parry asked Robert Brown to help with the botanical specimens but he was ill and this delayed the publication even longer.[8]

It was the Inuit and their domestic lives, including their igloos, kayaks and umiaks, their dog-drawn sledges and the ingenious hunting and fishing techniques that became the major focus in this book. Lyon's efforts to compile a phonetic vocabulary of the most common words used by the Inuit of the Melville Peninsula was unique, but sadly, went largely unnoticed. Parry was encouraged, even though it represented less than the money he had been paid for his first journal, with the offer of 1,200 guineas (£75,000) from John Murray.

The two long winters had been spent mainly on the surveys to the north-eastern end of Repulse Bay, and central parts of the Melville Peninsula. They had travelled north as far as the *Hecla* and *Fury* Straits from where it had seemed a sea route was open to the west, navigable when the ice cleared. Their sledging expeditions had mapped most of the area on the eastern tip of North America, described as a 'bastion', and had eliminated the possibility of a channel through the western or northern sides of Hudson's Bay. These discoveries were geographically important but had little of the excitement and suspense that the winter on Melville Island, and on reaching the farthest point west to date, had created.

Parry summed up his voyage both privately and publicly as 'two successive winters of long and tedious confinement'.[9] Consoling himself somewhat he wrote a thoughtful but sad sentence in one letter to Barrow and Melville and another to brother Charles, that they had 'narrowed the ground by having proved, at least what was *not* to be done.'

Before arriving back in Scotland, Parry had decided that he should return again the following year searching for the elusive link between the western end of the *Hecla* and *Fury* Straits, which he judged led into the southern extremity of the Prince Regent Inlet.

Jane Griffin, later to become Franklin's second wife, noticed during one party that Parry 'exhibited traces of heartfelt and recent suffering and seemed to be going *against* his inclination'. This is uncorroborated by any letters to friends or family and, since he believed that the next expedition was more likely to discover a final route, it would be entirely appropriate for him to finish the project he had pioneered from the start.

Proof of this determination was evident when, unexpectedly, he was offered the appointment of assistant hydrographer at the personal request of Viscount Melville, the First Sea Lord. He accepted on condition that he was allowed to continue with *his* plans, for finding the passage. Melville assured him that it would not interfere in any way and would provide extra income and greater security for that time when his exploring career was over. The charts he had prepared in the past and recently in Repulse Bay was the obvious explanation for this offer although owing to the usual parsimony of the Admiralty, he would not be paid any part of his £500 salary until he returned from the expedition. Barrow too was in favour of this appointment, describing it as: 'his sheet anchor for the future'.[10]

During the winter months of 1823/24 temperatures throughout Britain were abnormally low. Parry had little respite from the freezing conditions in the Arctic while at Deptford, once more supervising the fitting out of *Hecla* and *Fury*. Franklin was also in the dockyard organising the construction of three special boats for an expedition down the MacKenzie River to the Canadian Arctic to the west of the Coppermine. Franklin was now married to Eleanor Porden, and they were expecting their first child, born on the 3 June 1824.

The two men must have discussed the probability that should a navigable sea route be discovered it could only remain open for a few weeks in any normal year. It would be dangerous to navigate through due to the constantly shifting and unpredictable types of ice, frostbite among crews, extremes of wind interspersed by dense fog and heavy snow storms – and there were still unknown areas of the Polar Sea full of uncharted shoals and rocks. These were difficulties which trading ships would be ill-equipped to overcome and return with any worthwhile profit. The Government expeditions had been launched with the desire for geopolitical and scientific advantages not mercantile gains; the commercial aspect was now never mentioned.

The real intention was to secure scientific information and to claim a strategic position before rival nations, especially the Russians, could occupy the territory. The Royal Navy and its officers were entirely willing to emulate the efforts made by Captain James Cook, who had tried to find the passage from the western side through the Straits of Bering in 1778. But his attempt was made far too late in the season and ran into heavy pack ice. Neither of his two ships had been strengthened or prepared for this task.

Parry and Franklin knew their careers would be swiftly improved if a passage was found. This grail was therefore important for both men, but it was not, strangely, the sole reason for years of physical hardship and mental stress. They, like most Arctic veterans, became involved emotionally. Polar regions were a strange addiction and peculiarly attractive to devout Christians. Silence and space, mysterious and unpolluted, they were in stark contrast to noisy, foul-smelling, disease-infected cities and towns in industrial Britain in the early nineteenth century. For some they gave a distinct message of a transcendental nature relating to their own and mankind's destiny.

The colours deep inside icebergs, wonderful prisms of light reflected from glacial seracs and ice-tipped mountain peaks, and the shapes of animals and angels in ice floes. The *parheliae* or false suns, and in darkness the glimmering brilliance of stars and the wide spread Aurora Borealae appearing and disappearing without warning. To people with even a slight feeling for religion these could be taken as clear evidence of some divine presence. The Arctic journals viewed collectively mention often a sense of some heavenly guidance usually described as 'providence', and a feeling of a Divine protection (See Fig. 11). The prismatic cross is one example. This strange atmospheric effect appeared in front of the moon and was originally drawn by Captain Belcher RN during his searches for Franklin in 1854.

Franklin, like Parry, organised every aspect of his expeditions with complete attention to detail. Despite his preparation the 'awfulness' of the wasteland he passed through and their survival were, he recognised, chiefly due to God's loving protection. In most respects he resembled the biblical pilgrim.[11] He stood by his conviction that whatever the dangers and difficulties, the marvels and dangers in nature were the work of God alone and He would be their constant companion and guardian.

Parry was an evangelical Christian even before his expeditions started and believed that his mission in life was not only discovery, but the improvement of the lives and conditions of his sailors. Even on his encounters with Inuit people he was tempted to evangelise and convert them to Christianity.

In letters written to his mother and brother Charles he continued to express regret for having been away at the time of Caleb's death and of being unable to help with the many problems stemming from the tangled financial affairs from his estate.[12] Charles struggled on with these complicated problems for many years.

The arrangements for their mother Sarah, and the six children of their late sister Emma Eardley-Wilmot, were still concerns for Parry, as well as his headstrong young sister Caroline. He expressed doubts about her intended marriage to Joseph Martineau, a partner in Samuel Whitbread's Brewery and a Huguenot, although he had been given 'no details'. Parry asked for an urgent meeting with Martineau, which did not happen; fortunately for them all the situation was resolved by their mother, who gave her full consent.[13] Parry and Martineau eventually became close friends.

Parry looked for many ways in which to fulfil family responsibilities and wrote a series of letters to Charles on different subjects, but his desire to be of help was outweighed by the considerable time needed for finding suitable crews, fitting out the ships with some of his latest ideas and constantly controlling and inspecting the long list of supplies they were loading.[14]

Parry chose his officers from the same group he knew well and who were Arctic veterans. They including Joseph Sherer; Henry Foster, an astronomer-surveyor recently elected as a new member of the Royal Society; Henry Hoppner, his second in command; James Clark Ross once more; and the ever dependable William Hooper as purser. Four experienced 'ice masters' were signed on, Allison and Crawford who had fifty years of whaling experience between them, and Fife and Alexander. There were eight young midshipmen in all including Horatio Nelson Head an artist, and Francis Rawdon. The Reverend George Fisher went along as chaplain and additional scientist.

The orders from the Admiralty were straightforward and less insistent on the purely scientific element. They were to re-enter Lancaster Sound and explore down the Prince Regent Inlet to its furthest extent. The route he had briefly begun in 1819 which ended after only sixty miles due to thick ice and persistent fog. If conditions allowed they were to pursue *all* promising routes leading west, and, if any passage was discovered, then continue to the Bering Straits.

They were to rendezvous with John Franklin, George Back, John Richardson and fourteen sailors paddling down the Mackenzie River, while yet another expedition was preparing to sail eastwards from the Bering Straits, collect Franklin's men if they reached them and return to Britain via the Pacific. This expedition on a small sloop, HMS *Blossom*, was commanded by Lieutenant Frederick Beechey who had been on *Hecla* in 1819.

Another ill-conceived and under-equipped expedition was offered to George Lyon using the unseaworthy, unsuitable, and by now very elderly ship the *Griper*, which had been so 'lubberly' on Parry's first voyage. Lyon was instructed to cross the Atlantic, land on the Melville Peninsula close to *Winter Island*, travel across to its western side using dog-hauled sledges provided by the local Inuit, and find the place marked by Franklin in 1819 called Point Turnagain. This would join up the map between Melville Peninsula and that position on the Canadian Arctic coastline. It would be a return journey of over 1,200 miles.

These attempts on the passage from both directions were not undertaken without considerable expense including four ships, over 100 men and officers with all the equipment and supplies needed for several years exploration. It could have cost at least £35,000 (£2,500,000) to the British taxpayer.

Barrow, determined to locate the passage before Parliamentary and public patience ran out, made sure that the Admiralty instructions to every commander put geographical discovery ahead of scientific experiment. The scientific work was to be conducted when ships were in harbour or blocked by ice.

Hecla was dressed overall with flags and coloured lights for a special party and a concert attended by Madame Pasta, a leading opera performer of the day, a few days before departure. This was attended by many official dignitaries and several members of the Stanley family, including Isabella; it was the second occasion recorded at which Parry and Isabella had met.

Caroline and her new husband Joseph Martineau were also there and she wrote later to her mother Sarah that it was 'a beautiful sight and under other circumstances we should have thoroughly enjoyed it'.[15]

Fury and *Hecla* left the Thames on 19 May 1824. Franklin followed eight months later on 6 February 1825. His wife Eleanor, close to death from tuberculosis, insisted that her husband should leave and urged him not to wait saying, 'It would be better for me that you were gone.' She died six days afterwards.

From the start the Parry expedition had difficulties. Crossing Baffins Bay was once again hindered by thick middle ice stretching out over miles to the west and north in the Davis Straits followed by a long and severe gale from 1 August that nearly capsized *Hecla*. This was then replaced by days of thick fog. As a result the crossing lasted nearly to the end of the summer exploration season so that when finally they arrived at the entrance to Lancaster Sound they had already been away from Britain for three months.

Entering the Prince Regent Inlet round Cape York the winter ice was already starting to block the direct route due south as it had done five years earlier. Without sufficient time to make progress to the south and to find a suitable winter harbour on the American side, Parry decided to remain at Port Bowen on the north eastern shore. This was a safe but dreary place already discovered by him in 1819. This would now be his fourth winter season in the Arctic with only two winters elsewhere – and having lived through the freezing conditions in London.

Hooper, on *Hecla*, present on both the earlier voyages and described by Parry as his 'right hand', proposed in a resolutely cheerful and confident way that they were 'at the threshold of the hinge on which our future operations will turn'.[16] His optimism was not matched by that of his commander, who no doubt, remembering earlier disappointments, could find precious little encouragement in their position.

Parry wrote that 'these disappointments of the past season had served to temper the somewhat extravagant enthusiasm of those who not having been on the former voyage, could see no difficulties in the way of immediate progress'. He was still uncertain over tidal movement in Prince Regent Inlet, which if absent might reveal it to be another cul de sac resulting in nothing new being found to the south when spring came. The Admiralty instructions were, however, explicit. He must explore the entire Inlet to its furthest limits and not turn back until he had done so.

Despite these misgivings and suffering from rheumatism in his groin and thighs, Parry did his utmost to show confidence and maintain morale.

Several shore observatories were built which, according to Hooper, resembled a small village. These were used for the usual tests of instruments and to record compass variations. Parry and Henry Foster calculated that between the true and magnetic poles' position, a difference of 9 deg. from 114° to 123° now existed, far more than when last measured in 1819. From this fact they concluded that the magnetic pole was constantly moving and prepared statistical evidence as proof for the Royal Society.

The entertainments provided during the previous voyages were, Parry admitted, 'almost worn threadbare', nor were there any friendly Inuit in this gloomy place to divert, amuse and interest them. The ship's newspaper was not produced but, due to Hoppner and Hooper's ingenuity, several costumed masquerades were performed by the officers, including one in which Parry took part playing a well known old marine character frequenting Portsmouth dockyard. For this part he had face makeup, a wooden leg and a tattered costume. Naval regulations insisted upon tight discipline at all times and acts that reduced officers' authority among the lower deck were frowned upon. It was not supposed that officers would act on any stage.

Parry wrote in his voyage account, possibly as a minor precaution, 'no instance occurred that would weaken the respect of the men towards their superiors'.

Evening classes and Bible readings were given by Hooper and became part of the daily routine, often with the entire ship's company taking part. Parry commented 'such a scene of quiet rational occupation as I never before witnessed on a ship'. He felt sure that their study contributed to 'sober cheerfulness, uninterrupted good order and an extraordinary level of health all through the winter'.

A new barrel organ was played at church services and at a celebration on Christmas Day, as well as the inaugural Grand Carnival Day. Despite these efforts the winter, unlike that spent at Igloolik, felt interminable, and once again the ice inshore remained well into July 1825.

During the long tedious weeks at Port Bowen Parry's journal becomes increasingly pessimistic: 'all is dreary monotonous whiteness – not merely for days or weeks, but for more than half a year'. His descriptions include phrases such as 'inanimate stillness', and 'motionless torpor' to describe the bleak landscape devoid of animals and birds.[17] He was deteriorating physically and Hooper expressed concern over his loss of appetite, weight, severe headaches and insomnia. One modern diagnosis could be that he had symptoms of seasonal affective disorder, but this was not recognised at the time. His rheumatism affected his physical ability so much that instead of leading expeditions as usual when the spring started, he had to prepare himself and rest, ready for the next and most difficult part of the expedition.[18]

He delegated most of his work to others, something he had refused to countenance earlier. The repetitive duties of monitoring compass readings for weeks on end and the frustration of James Ross and several of the officers at the utter tedium of their surroundings suggested to him an expedition that would map the coast between Port Bowen and the tip of Cape York, 100 miles away. He did not offer to go and Ross reported later that it was 'all through dreary and empty country where hunting had little success'.

Morale among the crew might have deteriorated much further over the long winter but for the improvements Parry had made to their clothing and footwear. They had been given pea jackets of box cloth very closely woven and almost impervious to wind and rain, and special lined boots of soft leather for warmth. They were testing for the first time a brand new waterproof material supplied by a Mr Macintosh of Glasgow.

There were also different canned food recipes and supplies of pemmican, taken at Franklin's suggestion. These changes to diet were popular and offered more variety, carbohydrates for bulk and sufficient protein to maintain strength during sledging trips. No one had yet shown any signs of scurvy.

The Sylvester stoves which were placed on the orlop deck, the lowest part of the ship, burnt continuously, sending warm water through pipes to most places on board including his own cabin, which, Parry claimed, never fell below a temperature of 11°C.

After laboriously sawing their way free of the surrounding ice they came to clear water and sails were hoisted on 20 July 1824. The plan was to sail south over to the western side of the Inlet searching for open channels of water. By moving south it was likely they would find thinner ice, but the rock-strewn shoreline on that side was still surrounded with ice floes and overshadowed by immense cliffs, 450ft high. On 1 August, having gained sixty miles, they were forced on shore by some moving pack ice and while trying to refloat *Fury* she was badly damaged. The only solution was to beach her and undertake repairs as best they could. (See Fig. 14.) The *Fury* was on a narrow beach below the towering cliff and in a dreadfully exposed position, shown in the compelling sketch by Midshipman Horatio Head (Fig. 14.)

The stores were all unloaded onto the beach and *Fury* heaved out by the crew to lie on her undamaged side. Her sailors were transferred into *Hecla*, but bad weather then intervened and the improvised dockyard was destroyed by strong winds.

According to the notebook entry of Midshipman Francis Crozier on *Hecla* for 19 August 1825, *Hecla* was now in danger too and Parry addressed all the men on her deck warning that they must prepare to abandon *Fury* if *Hecla* 'is no longer kept safe should the ice come in'.[19] They would still try to repair the damaged ship but must prepare to sail at a moment's notice – and the repair work needed to be completed the same night! Midshipman Crozier continued: 'I need scarcely add the ship *Hecla* was ready for sea at the expected time'.

On the next high tide yet more ice floes came inshore and they had to rapidly move *Hecla* out of danger. Parry then rowed over with Hoppner to *Fury* and found she had been damaged beyond repair and now had a broken keel. Both realised that repairs would be impossible and she was abandoned on 25 August 1825.

Parry's reflection on this disaster was pragmatic He wrote, 'a vessel of whatever magnitude, or whatever strength is little better than a nutshell, when obliged to withstand the pressure of the unyielding ground on one side and a moving body of ice on the other'.[20] Hooper's equally sanguine comment was: 'the only real cause for wonder is our long exemption from such a disaster'.

None of them could anticipate how their 'disaster' would forestall a far greater one a few years later, and that the beneficiary would be none other than John Ross. He entered Prince Regent Inlet intent on salvaging his damaged reputation and to surpass Parry's record but was trapped for four consecutive winters by the ice. The supplies of food, coal, small boats and useful chandlery left on Fury beach provided all the resources from which to construct a large wooden hut and remain alive during the winter of 1832/33 before being rescued by the whaler *Isabella*.[21]

A crowded *Hecla* came alongside in Peterhead Harbour on 12 October 1825, having lost her companion ship but with only two crew members dead. One from an accident and the other from an incurable disease.

Parry was now less certain that a route would be discovered through Prince Regent Inlet and these views were confirmed ten years later by Ross' attempt; no new evidence had been gathered either way but the ice had twice shown him that navigation was nearly impossible at any time of the year. This was Parry's final attempt to locate the passage and his opinion now was that even if discovered, two or possibly three summer seasons would be required to fully complete the transit.[22]

News that none of the other expeditions had succeeded was of little consolation while his friend George Lyon, commanding the old and unseaworthy *Griper*, had fared considerably worse, having been unable to get even so far as Hudson's Bay.

John Barrow still maintained that discovering a passage would be worthwhile despite the mounting cost to the nation. He was still certain that the Royal Navy's participation would become increasingly important to many new sciences such as geology, minerology, and even broader studies now described as oceanography, as well as astronomy and hydrography. But apart from a very few individuals the large amounts of new scientific data, most of which had not been consolidated, was never seen and simply stored away.

Only a few hundred miles of coast remained unexplored and since Parry's opinion remained positive, albeit dubious about the time it would take to find the route, Barrow as obstinate as ever was still prepared to continue. The last major expedition which set out twenty years later resulted in the death of John Franklin with all his 129 men.

Chapter 10

Marriage and the Stanleys

The first of two letters written by Parry in England was to Sarah his mother, posted while travelling down to London on 16 October 1825, telling her that he was safe and well, and another to Charles given to Hooper who delivered it on his way home to Clifton, north of Bristol. He begged for news of the family and especially Charles, who he had learned had been ill with asthma.

In the letter to his mother he asks that each member of the family adds a line to her reply, which should be sent to Spring Gardens Hotel in London where he would be staying, rather than to Barrow at the Admiralty.[1] Another letter to Charles from London on 22 October explained that he would have only forty-eight hours at Bath since he was due in Sheerness for Hoppner's court martial on 24 October (the commander of any Royal Navy ship that had been lost was subject to a court martial to determine the precise reasons and the actions taken to save the vessel). As Hoppner's superior, Parry was witness to the loss of *Fury* but surprisingly was also appointed a member of the court.

Sarah was at Wickham Rectory near Bishopstoke in Hampshire, with the Garnier family. Their son, Thomas Garnier, had married Parry's sister Maria, who suggested to him that Sarah needed extra care.[2] He asked Charles to write to him at Sheerness 'to ask questions in your letter', suggesting that he did not know what kind of news was wanted from him, nor how much. The recent voyage required his considered account of the problems they had overcome and the decisions he had made; this took most of his time and he was frustrated by the delay in returning home to greet his family, whom he had missed badly.

At the court martial Hoppner, as commander of the *Fury*, was completely exonerated and pronounced 'blameless'. In his next letter to Sarah at Bishopstoke Parry wrote, 'the court martial sentence is not only satisfactory but flattering to all parties'. Charles got the same news in a letter sent down with Lieutenant Sherer, who would 'provide him at first hand, all the details, 'and only then could he go down to Bishopstoke for a brief stay of a day – a very short visit considering his long absence. He wanted to discuss with his mother his plan, which he considered 'most conducive to her own quiet and comfort'.[3] Parry wanted her and his unmarried sister Gertrude to stay in a respectable London hotel so that they could be closer while he was working back at the Admiralty. Charles, however, resisted his idea and proposed that she went back again to Bath where most of her friends and immediate family lived. Parry, despite his disappointment, understood that it was best for their mother to decide what she wished to do. Writing to Gertrude in a letter dated 25 October he announced that he was, 'much hurried but remarkably well'. The truth though was different, as he was having frequent bad headaches for which he had taken special drugs, severe rheumatism in his back and groin, and symptoms of depression. The next day he wrote an eight page letter to his brother which explained most of his worries and irritations. He began by telling him the history of his encounter with Miss Browne, the true state of his health, his low income and how he had 'always felt the desire to be attached somewhere'.[4]

He was deeply angered at the treatment of his friend George Lyon by the Admiralty Board. His career with the Navy had come to an abrupt end after the elderly and unseaworthy *Griper*, battered by Atlantic storms, returned to Britain without her anchors and only just afloat. She had never reached the Melville Peninsula.

This had all been blamed on Lyon's captaincy but Parry accused the Board of blaming everyone but themselves and that, 'it is the Admiralty with whom the principal original and most glaring fault lies'. He asks Charles to burn this passionate letter, 'with its secret and treasonable matter'.

Parry told Maxwell and close friends, including Edward Penrhyn, Edward Hankinson and Reverend Edward Stanley, that his poor health was due to the 'irksomeness of sedentary life', and made known his strong desire to find a mate. His depressed state did nothing to prevent him from trying harder to get Charles' agreement to their mother's removal to London, saying that he would now remain in town for a good length of time, 'no more polar expeditions certainly'.[5] But Charles continued to object.

Parry's long relationship with the Cresswell family, whom he had known since childhood, now helped him recover as several invitations came to visit their home near Kings Lynn. At one weekend party Catherine Hankinson, then a girl of 18 and later to become Parry's second wife, described him favourably in a letter to her brother Edward Hankinson.

> He is I think, the most good tempered man I ever saw; he is very musical, and sang a great deal to us ... He was extremely pressing that every-one should sing and it was by only the greatest entreaty that that I could prevent him from going down on his knees to me. I wish you had been there, I am sure you would have enjoyed it; Captain Parry is so very entertaining.[6]

In November Parry found lodgings in Jermyn Street, close to his office in Somerset House. He was confirmed as hydrographer to the Royal Navy on 11 November 1825, but it was a position he found both irksome and terribly unrewarding due to the shortage of staff, a huge backlog of work, and the uneasy relationship between John Barrow and John Wilson Croker, the First Secretary to the Board of the Admiralty who reported, like Barrow, directly to Viscount Melville.

Croker was a Member of Parliament representing Admiralty interests in the Commons, an antiquarian and used to manipulating and partly managing and allocating its budgets for more than twenty years. He considered the Arctic expeditions an unnecessary expense and consistently said so. To Parry's discomfort he was also reluctant to increase his small hydrographer's departmental budget, resulting in too few clerks to draft and copy the flood of important new charts originating from surveys across many oceans.

London gave Parry a welcome opportunity for distraction since the London Season, beginning in April, included many private dinners, dances, concerts and theatrical performances. Being a national celebrity, he was greatly in demand. He renewed his friendship with Lord Stanley's daughters, particularly the youngest, Isabella, now 24 years old and the acknowledged beauty of her family. Parry was 35 and had physically aged during his last two expeditions but despite several ailments, remained a vigorous and handsome man.

By the middle of May he had decided to write to her father, Sir John Stanley, declaring his feelings. This was a letter in which he stressed a conviction that she possessed the most important qualities he wished for in a wife.

> I perceive indeed so much congeniality in our views and sentiments on all essential points, that I cannot but consider my own future peace of mind likely to become deeply involved and at least an injustice done to your excellent daughter, if I longer hesitate to communicate with you on this subject.[7]

Parry knew that Lady Maria did not favour his courtship of their fourth daughter, and so decided to appeal directly to Sir John. Adhering strictly to the formal practices of the time, Parry needed to win an important ally on his behalf. Sir John was a mild-mannered, unambitious member of the aristocracy, used to a life of leisure and ease but had travelled to Iceland when young, and been elected into the Royal Society, which stood firmly in Parry's favour.

He was interested in topics involving the sciences, and was aware of the many instrument experiments and the geographic discoveries made by Parry and the Royal Society nominees. Parry declared his income, which was not large being slightly under £1,000 pa, (£67,000) including his captain's half pay of £191 pa (£13,000) and hydrographer's pay of £500 (£34,000). The rest came from rentals on four houses in Bath. His letter ended, 'In any case, and with whatever pain and sacrifice to the feelings I have lately ventured to indulge, I promise implicitly to abide by your decision.'

He was summoned to the Stanleys' London town house at 6, Audley Square, where it became clear that Lady Maria, the matriarch, was going to conduct a gruelling inquisition of her daughter's suitor. The differences between them over religion, the Quaker movement, especially the Fry family, arose rapidly and Parry wrote afterwards to sister Gertrude that:

> In a long and not very agreeable conversation with Lady M. S. she has been trying to scold me out of my religious sentiments and informing me of her utter abhorrence of Mrs Fry and all her Saints; this I understand to be deliberately provocative because my friendship with the Fry household is mutually positive.[8] Isabella has for some time past been acquainted with my views and I am sure she approves my adhering to my principles and *being consistent*.[9] I trust in God that whatever happens I may never give *him* up even if I have to give up Isabella herself.

It was a defiant stand.

He had recently become involved with several religious societies, including the Church Missionary Society. He had no illusions about the obstacles Lady Maria was capable of putting in his path and for that reason told Gertrude, to 'use this letter as you see best – I recommend rather checking Mother's dearest hopes, because I begin to check mine. All may yet be well – but it may not.' Several other difficulties lay in his path, of which one was the recurring ghost of Miss Browne, a distant relation of the Stanleys, and whose mother complained that Parry had jilted her daughter.

Another was the rejection of an earlier suitor by Lady Maria, but for some unspecified reason nobody could or would divulge. The relevant Stanley correspondence has been lost, preventing anyone knowing the reason, but one most likely cause was his class, income, occupation, age or all of these combined. Parry, at 35, was not too old to marry Isabella, but the Stanleys were part of a very old aristocratic line and held themselves equal to the senior family members such as Lords Derby and Sheffield. They had no wish to encourage suitors who could lower the prestige of the junior branch.[10]

Partly due to his lasting friendship with Reverend and Mrs Edward Stanley, he was invited to spend the summer of 1826 at their rectory near to the entrance of the large Stanley estate at Alderley Park in Cheshire. After two months of miserable separation, Isabella and he were able to meet in the rectory and walk in its grounds. The objections by Lady Maria to their engagement slowly began to disappear and Isabella wrote to her sisters, 'Mama is changed. Papa likes him and I have nothing now more to wish.'[11]

From the rectory Parry began to explore the rolling parkland with its beautiful beech woods and long views from Alderley Edge over the Cheshire plain below. The Stanley house was neither beautiful nor grand, once having been a large Hall that had burned down and was then rebuilt in stages by Sir John's grandfather. What it lacked in beauty, however, it made up for in charm.

The six Stanley children were either very dark or very fair. All were affectionate, disputatious, and lively. The unmarried daughters, Rianette, Lucy and Louisa were close in age and inseparable. Edward John, the oldest son and heir, and his twin brother William, were headstrong but in other respects quite unlike one another. Isabella's choice of husband was of the greatest importance to each one and this became a time consuming trial for Parry who wanted to win their individual as well as their collective approval.

Isabella suffered badly being of a romantic and gentle nature. She was incapable of mentally blotting out all the strong opinions volubly expressed by her family – some becoming storms that lasted for days and involving the entire household. She was the complete opposite of her mother who was a forcible, forthright, highly rational and domineering personality. Lady Maria the eldest daughter of Lord Sheffield could argue, it was said, with anybody about anything and usually did so, and it took courage and determination on Isabella's part to keep her engagement alive.

In mid-August only days before the family were due to move to their estate at Penrhos on Anglesey, a wedding date and place was finally decided: 23 October, at Alderley Park. Parry would remain in London meanwhile, working at the Admiralty and beginning preparations for a new expedition, something he had previously told Charles that he would no longer consider.

On 23 October a cloudy and rainy day, the wedding party met at Alderley Park for breakfast. It was a large gathering of the Stanley and Sheffield relations with a few of Isabella and Parry's closest friends. There were six bridesmaids including all the Stanley daughters but no one from his own family was present. For Sarah this was understandable, she was 71 and in poor health, but that none of his sisters, his brother or his uncle Sir Benjamin Hobhouse were there seems odd, since both the latter were also members and Fellows of the Royal Society and well acquainted with Sir John. Among themselves they may have decided not to attend but it is likely that the guest list never included them at all. Parry certainly did not appear to argue in their favour with his future mother-in-law and may have deferred to her wishes. No letters or journal entries by either Parry or Isabella explain the reasons, but irrespective of how this came about it did not create any Parry family friction. His marriage into an important aristocratic family and his new place in society had risen well above theirs.

His acquaintances included Lords Bexley, Morpeth, Prudhoe, Beresford, Viscount Melville and Lord Seymour, as well as recognition by the Duke of Clarence and Prince Leopold.[12] No detachment from his family's news or affairs is apparent in his letters, although sometimes he is self-consciously proud to list all the names of the titled men he has recently met or dined with. The family's replies are no longer available, but the frequency and number of his letters sent after his marriage are shorter even if no less frequent than before. Undeniably, his life and theirs were now wider apart.

Before his wedding day Parry admitted to being 'dreadfully frightened' and Reverend Edward Stanley, who conducted the marriage service, had difficulty in delivering his words. A subsequent letter to his mother giving an account of the wedding mentions: 'how many and how dear were the relations *she* was leaving and leaving for me'. His own letter was appended by a note that Lady Maria Stanley had generously 'offered to write you a short account of the wedding tomorrow, which will be sent by the Birmingham mail'.[13] The bride and groom went away from the wedding in a carriage given them by Sarah as her wedding present, to a house in Sheen outside Richmond Park, belonging to Isabella's cousins, the Penrhyns. They were now alone for the first time for many months.

From Sheen Isabella sent a letter to her sister Louisa.

> You cannot conceive how comfortable I am altogether. Living alone thus with him and entirely with one person, one's own husband – does indeed draw us close to each other and bring out every feeling of love and affection, of a kind deep and tender such as one has never felt before and yet with this feeling though he seems to possess all one's love – yet it does not prevent me from loving you *all* if possible dearer than ever. No, it is a feeling which I am sure enlarges the Heart and gives it only new powers, new energies without chilling or taking away any old feeling of affection. Firm and steady as a rock may I depend upon him whom I have chosen.
>
> There is nothing hidden or variable about him. He is everything open manly and true and with all the tenderness and feeling a woman could desire.[14]

When his two weeks of leave were over they moved into the Stanley town house in Audley Square where Isabella had to hire a cook, a footman, and make the house ready for the round of entertaining expected for their friends and relations including the Hobhouses, his uncle and aunt, and the Carpenters who were hers. They visited the Martineaus at Stamford Hill who had given Parry a second home during his brief periods ashore over the previous eight years. During the second of those visits hosted by Caroline, the Martineaus gave Isabella a true account of Miss Browne's dubious history. Isabella said afterwards that, 'she wished that if Mama could have heard it, she would never have taken her part'.

Before marrying Isabella, Parry had been searching for any possible way of escaping the drudgery of the hydrographers office, and had consequently proposed an expedition to locate the North Pole despite his statement given in writing to Charles that, 'I should like to be quiet after eight years of continual hazarding both physical and medical even though *I do not expect it*.' In fact it was on his own initiative that he approached Viscount Melville in April 1826, having been handed a copy of Franklin's identical proposal by John Barrow.

Parry had visited only parts of the eastern side of the Greenland Sea during his short wartime patrols around Spitzbergen. Despite this the proposal was supported by Melville, albeit reluctantly, as well by Sir Humphry Davy, President of the Royal Society and the Board of Longitude. This expedition, he had decided, would dispel any disappointment at failing to find the passage. There was now a distinct possibility that John Franklin might do so, as news from a Mr Kendall at the Admiralty was that Franklin and his party had reached the Arctic Sea and could reach Bering's Straits later that year. They had completed the exploration of the McKenzie River only eighteen months after leaving the Thames.[15]

Parry's proposal to Melville was that it would take little time to reach the Pole and presented far fewer difficulties than any of his earlier expeditions. He was absolutely confident of his ability to carry it through successfully. He did not think to ask himself the most important question of whether was he the best person to make this attempt. Franklin had petitioned Barrow five years earlier and prepared a detailed plan for the project. Parry's willingness to step into his friend's shoes, even though Franklin was several thousand miles away, suggests a different side to his character both opportunistic and ambitious. His earlier expeditions had been made on ships, whereas Franklin had faced and overcome many unusual ice conditions on land very different to those at sea. He had devised lightweight equipment such as tents and clothing for this task and led cross-country marches through broken terrain, frozen lakes and river torrents using canoes and portable boats.

The expedition to the North Pole was given priority by the Royal Society since it would provide a detailed record of magnetic variations at the furthest point in the northern hemisphere. It would also help decide on the true shape of the earth using Kater's pendulum instrument and what the area surrounding the Pole looked like. Was it land or water? These were their chief objectives but, as before, they would take many more weighty instruments for measuring ice thickness and salinity, wind speeds and general meteorological conditions, astronomy and hourly temperatures at the Pole. They were ordered to measure how far any flat land extended around the Pole, but no single method of calculating this area was proposed.[16]

If a Polar Sea existed it could then be claimed as another important British discovery.[17] Barrow hoped that the success of this new project could assuage increasing public doubt about the existence of the Northwest Passage after a decade of failures, including Parry's. The British public's consciousness could not easily separate the Northwest Passage from the North Pole, and the fact that a magnetic North Pole is different from its geographical version, was unimaginable. When the magnetic North Pole was pinpointed by James Clark Ross in 1834, the public – greatly encouraged by John Ross – thought it to be the true North Pole at N.90 deg. of latitude.

Many of the experienced Arctic officers were away from Britain taking part in Barrow's concentrated assault on the Passage, but there were several suitable officers available. Lieutenant James Clark Ross, veteran of Parry's three previous voyages, as well as Crozier, Hoppner and Hooper from the most recent voyage down Prince Regent Inlet. Both Ross and Lieutenant Francis Rawdon Crozier were experienced navigators, good leaders, resourceful survivors. They had become close both socially and professionally. Ross' acceptance of the new commission increased his own certainty that they could complete the task in one summer season. Neither had any appetite for months of privation in freezing darkness if isolated near the Pole. He explained this to Isabella and her family and that his absence would be six months at most. He also wrote to Sarah, then in Bishopstoke with her carer Jemima, along similar lines, 'few enterprises are so easily practicable. It is a most interesting and delightful one.'[18]

His sincere wish to settle down and enjoy married life was to be jeopardised in the short term for what he doubtless hoped would be a fitting conclusion to his remarkable expeditionary career. Success would inevitably lead to promotion with increased income, he would escape the tedium of the Hydrographers Office as well as the Government bounty of £5,000 (£450,000)for any ship getting within 1 degree of the Pole.

On top of these, money from yet another publishing contract with John Murray was offered.[19] The publisher had suggested the large sum of £1,500 (£90,000 today) for an omnibus edition of all his journals. He now seems to have placed any other difficulties to one side and uncharacteristically failed to adopt sound advice from experienced travellers to the polar ice cap.[20] Nor did he apply practices he had witnessed on the Melville Peninsula in terms of proper clothing, travel with dogs and sledges and most of all, footwear. There was a hastiness with the preparations, and too great a reliance on John Barrow's conviction about an open sea at the Pole, which led to unwise decisions. Viscount Melville was *much* more doubtful about his proposals, but Barrow and the Royal Society, devoted to completion of their global magnetic survey, got their own way. William Scoresby advised them to take light sledges pulled by trained dogs with skilled Norwegian drivers but he was ignored, as too were his recommendations that they must start in late April or very early in May at the latest, when the ice cap would be frozen hard and remain smooth. As summer advanced, he warned, the surface ice became slushy and by night would re-form as hard slippery hummocks submerged under pools of slushy rainwater, which made travel slow and difficult.

The *Hecla* was fitted out as the expeditionary ship and would leave England in April 1827, sailing first to Hammerfest in Norway to load eight reindeer. These would be used for hauling and carrying extra rations and when no longer useful for that purpose killed for food. Instead of using light sledges which Scoresby recommended, they built two large boats 70ft long and 20ft wide, weighing three quarters of a ton each, and fitted with steel runners. At one point Parry considered fitting wheels, but realised that these would dig into the surface and freeze on their axles. Instead of being on the edge of the icecap in early April as Scoresby had so strongly recommended he now decided to sail from Hammerfest to Spitzbergen to find a safe harbour for the *Hecla* as a base and sail both boats with their reindeer and men to the edge of the polar ice cap. He planned to start from that place on or by 10 June 1827.

During the months following his marriage Parry was away from home frequently, supervising work at Deptford on the expedition boats taken from Franklin's original designs, and working long hours at the Hydrography Office in Springfields. Eventually, on 13 November 1826, a new commissioning pennant was raised on *Hecla* with Isabella in attendance.[21] She was by now pregnant with their first child.

They both went down to Bath for his thirty-sixth birthday on 19 December and stayed for Christmas. Sarah had decided to live at Summerhill with Charles and his family, and Gertrude too. This was the first time Isabella had met his family and there was some nervousness on all sides.[22] In a letter to her sister Louisa, Isabella makes clear that the family viewed the Stanleys as far above their own station in life, writing:

> though they are in a set lower than ours and live in a quiet and different way, yet there is not the least particle of vulgarity about any of them. I see

plainly that they were rather afraid of my not liking them and that I should look down upon them, but they have quite lost that fear now.

The social distinctions are apparent to everyone but she is careful to say very tactfully: 'Mama would be quite satisfied to know and see how proud they are of having a Stanley belonging to them.' She formed a warm relationship with Sarah and explained to her sister that:

They do (I mean his family), rejoice in seeing how different he is to what they have ever seen before. He had certainly been a great source of anxiety and trouble to them as they knew he was not happy and always feared his meeting with disappointment and sorrow knowing what an effect it *always* had upon him.

These words and several other compliments about their education, connection to good society, and pleasanter manners than many more fashionable people, were no doubt designed to satisfy Lady Stanley, whom she well knew would read her letter.

During their stay Isabella was unwell and Charles, as a qualified doctor, was a welcome presence, whilst his wife gave sound advice and comfort. Isabella lay on a sofa for most of their visit but still received numbers of his friends and managed to attend a large dinner party hosted by Charles. Her condition did not improve however, and in January 1827, once more back in London, she miscarried.

As one means of reconciling the now weakened Isabella to his expedition, and to prove how safe and comfortable he would be during the time he was away, he arranged for them both to live aboard *Hecla* for a week. It would he hoped be reassuring for her to see him on the ship and to show how he would remain the same man she had married. Another visit on her own took place two months later in March when he was asked to preside at a court martial. During this second visit the Thames froze from shore to shore and around the ship and she was able to partly experience those freezing conditions he had survived for many months. She apparently enjoyed this experience and listened to the ice 'grate along the side of the ship'. She was deeply impressed by James Ross, whom she described as having 'much spirit and eagerness about him and that his whole heart and soul are in the present expedition'.[23] She had increased confidence in his ship, the competence of his sailors and the ways in which her husband's shipboard life matched their home life. He would, she believed, be a fine example to the men, supervising their spiritual as well as their physical welfare.

The success of this experiment was evident from her private journal entries. She understood that her full approval of the voyage was important to her husband and that she would be his spiritual and emotional support during his absence; along with most women, the role of the Arctic explorer was considered very manly, daring, morally congruent, and noble.[24] Isabella wrote: 'I would not recall you, your path leads to glory and honour and never would I turn you from that path when I know and feel it is the path that you ought to go.' In one further gesture of solidarity between them she agreed to let their favourite dog Fido go with him.[25]

Parry became even more optimistic with his plans as more stores were loaded, including some special new lamps run on spirits of wine, cooking pots which could be suspended over a new stove with seven wicks, wooden skis which no one had yet learnt to use, and foods including rations for an ounce of cocoa, ¾lb of best biscuit, the same amount of pemmican, plentiful lime juice and a gill of rum for every man each day. No tinned foods were taken because of their weight.

One unhelpful suggestion made by Sir Robert Jeppings of the Navy Board was that they should take velocipedes! This Parry knew was ridiculous and 'had not thought it worthwhile to get them'. He wrote to Charles explaining what special scientific equipment they needed and how they would establish their arrival at the true Pole by observation of the sun's altitude, if the sun were visible. He explained how the chronometers would be set and the distinction between the true and the magnetic poles and the calculations to be used. Towards the end he asks Charles in a jocular way, to be one of his two executors in case 'I am gored by Walruses', telling him that Edward Hanley, a friend, will be the other.[26] He wrote to Charles on 31 March saying that he hoped to be on the ice by 1 June, and that as he had no artist on board could he, Charles, produce some sketches from the 'scratches' they brought back. Edward Foster, his surveyor, would explore the north-east of Spitzbergen and hoped to meet an unknown race. He hoped this would not distract them from their main objective.[27]

At the end of March Sir John Stanley arranged a special dinner for *Hecla*'s crew, with their wives and children, to be held aboard ship. Sixty-five people came and Isabella was present, although now in the early days of another pregnancy and under doctor's instructions. She commented in her journal, 'a finer set of men I never saw; as our boat left the ship, the men stood in the rigging and along the side and cheered us away, which salute our boat returned, and the crew re-answered it. Never shall I forget this day.'[28]

On the last day of March 1827, *Hecla* was towed by a steam tug down to Northfleet to load instruments and be ready to catch wind for the voyage north. Isabella was not there to see them go. She had intended to return to Alderley from London but on doctor's advice had stayed with the Martineaus at Stamford Hill, where she confided in her journal that 'it' must now be her friend in his absence.

Chapter 11

Dark and Dismal Solitude

The late departure of *Hecla* created several unfortunate consequences and further delays before they finally reached the edge of the polar ice cap, reducing their chances of reaching the Pole in the time that Parry had allowed.

The first delay was in Norway where they had arranged to collect eight reindeer. After waiting three days at Hammerfest they went to fetch the animals themselves, but just before leaving the *Hecla* at Spitzbergen, Parry decided that controlling the animals, hauling their food along adding extra weight and possibly slowing them down, meant they would be of little useful purpose. He decided to leave all the deer on board *Hecla* as spare food.

The second arose from his original plan to join the summer whaling fleet about to move north to their fishing grounds and reach Spitzbergen in early May. The whaling captains all had good local knowledge and would help him choose a safe anchorage for *Hecla*, but the whaling fleet were late arriving that year and another week was lost; they finally left Hammerfest at the end of April.

Nearing Spitzbergen the ships ran into an extensive area of pack ice and were forced into its middle by strong winds and currents just as the *Trent* and *Dorothea* had been in 1818. Now forced northwards and eastwards along the coast their imprisonment lasted three weeks, and only then were they able to begin looking for a safe harbour. Another ten days passed before they found Treurenberg Bay on the northern side of West Spitzbergen. It was a full three weeks after the date intended and much later than Scoresby's recommended timing. They were approximately seventy miles from the edge of the polar icecap and one of their boats went over to inspect the ice and conditions generally which were pronounced 'discouraging', being described by Crozier very graphically as a 'Stone Mason's Yard'.

Lieutenants Foster and Crozier were left in charge of *Hecla* with a small crew, the reindeer and Fido the dog. Parry chose five of the strongest sailors for each of the two special boats fitted with steel runners. He would command one boat with Edward Bird while James Ross and Charles Beverly would take the other. The boats were named *Enterprise* and *Investigator* and flew silk ensigns made by Isabella. On the way over to the icecap stops were made at

Walden and Little Table Islands to leave food supplies for their return. Only seventy-one days' provisions went with them instead of the ninety-two he had planned but, despite this reduction, the weight for each man was 260lb and the boats lay dangerously low in the water. Finally, on 29 June – just over 100 miles from where *Hecla* was anchored – the edge of the ice cap was reached at 81 deg. 12' 51" N.

The two boats were covered with leather and oil cloth and were flat bottomed, which made them awkward to sail. Their size and weight meant they were also difficult to drag over the ice and because the Arctic summer was now well advanced they found the route filled with hummocky ice floes interspersed with narrow, shallow channels of water.[1] This forced them to haul out, frequently dragging the boats along to maintain a direct line. Just as Scoresby had predicted, the entire ice cap was becoming slushy and treacherous and as daytime temperatures rose they were obliged to travel at night when the ice hardened again and the fierce dazzle from the sun reflected off the ice was reduced.

Progress was slow from the start, barely three miles were accomplished most days. Ice conditions, which Parry and Ross had anticipated as broken and difficult, were expected on the edge of the 'main ice', but did not improve as they struggled northwards. The perpetual daylight now created disorientation and most of the group did not know what time of day or night it was. Parry wrote:

> This travelling by night and sleeping by day so completely inverted the natural order of things, that it was difficult to persuade oneself of the reality. Even the officers and myself who were all furnished with pocket chronometers, could not always bear in mind at what part of the twenty-four hours we had arrived; and there were several of the men who declared, and I believe truly, that they never knew night from day during the whole excursion.[2]

After a week of strenuous marches they were still only making three or four miles daily and their clothing and boots were permanently wet. Even though they rested in dry clothes, they started off again in the evening wearing half wet apparel from the night because there was no way of drying clothes completely.

> When we rose in the evening we commenced our day with prayers, after which we took off our fur sleeping dresses and put on those for travelling, the former being made of camblet a mixture of woven goats hair and silk, or camlot, lined with racoon skin and the latter being a strong blue box cloth similar to serge. We made a point of always putting on the same stockings and boots for travelling in whether they had dried during the

day or not and I believe it was only in five or six instances at the most, that they were not either still wet or hard frozen. This, indeed, was of no consequence beyond the first putting them on in this state, as they were sure to be thoroughly wet in a quarter of an hour after commencing our journey.

After breakfasting on warm cocoa and biscuit they marched and hauled for five or six hours before stopping for another meal and afterwards continued for another five hours. Their daily rations amounted to 10oz; less per man than that taken by Scott's expedition to the South Pole.

Each morning they halted and looked for a level piece of ice, pulled both boats together and rigged the sails as protective awnings with a single entrance at the bow. The men now changed into dry stockings and fur boots and went about repairing any damage to the boats after which they smoked a pipe. Parry, who did not smoke, allowed himself a refreshing splash of eau de cologne. A watch was kept in case of polar bears with everyone taking hourly turns and the bugle was sounded when the night's trek was to begin again.[3] Parry had never before experienced ice conditions like those on the polar ice cap. The complexity of variety and combinations in the ice were extraordinarily difficult to climb over. Sometimes there were high pressure ridges, sometimes mollochs or thin and decayed old ice on which snow lay deep enough to sink them above the knees, while others had 'holes quite through it in many parts which the smallest motion among the surrounding masses might have broken instantly into pieces'.[4] Not infrequently they took two hours to gain 100 yards, despite using all the men to haul a single boat over one short stretch unloaded, and then having to re-load it again afterwards. Six miles progress northwards was a very hard day's work but was considered good, although, as Parry himself admitted, it was less than half the daily distance intended. Rain fell on one occasion for thirty hours continuously. It covered the surface of the ice with pools the colour of green velvet, beautiful to look at but tightly packed with needle-sharp crystals resembling upright pen knives. As they tried to walk through the pools these cut through boots and socks and inflicted nasty flesh wounds which were hard to treat. The men carrying heavy loads were barely able to balance and cope with these obstacles and had insufficient food to maintain their full physical energy and body temperature.[5]

The amount of rain was more in total than Parry had recorded throughout his eight years of Arctic exploration, and was only equalled by the sun for hindrance and discomfort. The sun was so intense that it melted the tar caulking on their boats and created snow blindness from its powerful dazzle off the ice. Nothing came into view that relieved the monotony and dreariness. 'The eye wearied itself in vain to find any object but ice and sky to rest upon. From want of variety the most trifling circumstance such as the sighting of a single gull, engaged a more than ordinary share of our attention.'[6] The men laboured on with great cheerfulness and goodwill encouraged by the steadfast example of the officers.[7]

They were convinced that eventually they would reach a smooth and continuous plain leading up to the Pole, as described purely speculatively by earlier geographers. Finding no trace of this feature at noon on 17 July when the sky cleared and visibility improved, they stopped to rest, took sun sights, and completed the set of magnetic readings required for the Royal Society's isogonic survey. The quantity of work involved meant the officers and most of the men had to give up hours of badly needed sleep.

Five days later, on 22 July, Parry and Ross tried to reconcile the miles travelled against the calculations of each day's position and to their dismay realised they were being opposed by a southerly ice drift of four miles a day. The ice was carrying them backwards and halving the daily distance they made. On 26 July, when the calculations showed they had made only one mile to the north, Parry decided to tell his men the bad news and to turn back. Had they gone just one more degree north, to 83 degrees, they would have been entitled to the bounty of £1,000 (£68,000 today) offered by Parliament for reaching 82 deg. 40' and 23" north. Their total distance travelled was 660 miles with only 290 of these in the direction of the Pole.

It was a painful decision but he realised that the rations and their declining physical resources would be insufficient to complete the remaining distance and make the return journey. Ross injured his back while heaving one boat round an ice ridge and was crushed against an ice block. This made his decision easier as it looked possible that his spine had been damaged and that his mobility was now at risk. Starting the return journey immediately might give them time to regain *Hecla* for urgent medical treatment or surgery and a full recovery. Parry too was suffering again from severe rheumatism of the groin triggered by the cold and wet clothing all had to endure.

He now composed a draft letter to Isabella expressing his feelings about his decision and the nightmarish conditions that led up to it. He described with a heavy tone their gloomy surroundings, which he described as 'a dark and dismal solitude'. The letter was sent when they reached Orkney and was also a love letter and a passionate expression of his desire to be with her again. Absence from her 'was never again to be contemplated'. This letter was written on 27 July at the highest latitude they had reached, an achievement that remained unbeaten for the next fifty years.

After explaining the reason for his decision to the men he ordered a twenty-four hour rest to dry clothing and boots, to be followed by a special dinner. Despite his own despondency he was determined to keep morale high and encourage them all to believe in their success and a safe return. The new flags made for their arrival at the Pole, including the Union flag, were hoisted and many toasts – made with grog, rum and hot water – to the King, followed, at Ross' suggestion, by another to Mrs Parry. A sounding line 500ft in length was dropped through a hole in the ice but did not find the sea bed. One experiment which at least confirmed there was an Arctic Ocean under the icecap, although this was not at all what Barrow had supposed.

They now had to march 178 miles to reach the edge of the ice and sail another 100 miles over open sea to reach *Hecla*. A north wind blowing from behind as well as the same southerly ice drift provided welcome assistance, especially since food supplies were by now running very low. Some men had symptoms of scurvy and most were suffering from painful chilblains, numerous cuts and many open sores on their feet.

The return over the icecap took only two weeks, and at seven in the morning on 11 August they heard the swell and surge of the sea. They had a 'wildness in their looks', Parry wrote when they reached the edge of the ice at 81 deg. 34'N. One hour later they hoisted sails and set a course for Little Table Island, collected some stores they had left, much depleted by polar bears, and after several days of gale and snowstorms followed by tantalisingly calm windless weather, they reached *Hecla* at four o'clock on 21 August. They had been away sixty-one days in total.

The following week was used to regain strength and then they weighed anchor on the 28 August and set out for the Orkney Islands, which they reached three weeks later on 23 September. James Ross had made a surprisingly strong recovery and now remained on *Hecla* to bring her down south to the Thames.

Parry rapidly accepted an offer to return with his dog Fido and Lieutenant Beverley in a customs cutter just leaving Stromness for Inverness.

This was his final Arctic expedition during which he had once again shown his endurance, leadership, and bravery, as well as respect for his men's welfare. They had got nearer to the Pole than anyone before but the expedition failed mostly because too little attention had been paid to the wise practical advice given by William Scoresby Jnr. Scoresby had sailed around the polar ice cap, sledged across parts of it and even attempted a trek to the Pole himself. He had been proved entirely correct in proposing a start date one month sooner than they achieved and his advice that the heavy cumbersome boats should be left at the sea edge and replaced with strong, light sledges carrying canoes or umiaks pulled by dogs or by men on skis, was ignored. Their clothing should have been made of double layers of sealskin with warm and waterproof boots, and equipment adapted from Inuit inspired aids such as snow goggles. Franklin's original proposal, that Parry had rapidly appropriated, failed completely.

Their chances of success would have been been improved if such measures had been taken but few men could have lasted the 600 mile distance each way over tumbled floes, pressure ridges, decaying ice, sections of half melted water full of needle-like ice, and the southerly drift. Conditions on this occasion were so bad that even with modern clothing and equipment it was at the limit of men's endurance. The south-running current they had encountered did suggest another, and ingenious, means of reaching the North Pole. It was eventually used by the Norwegian explorer Frederick Nansen in his ship *Fran*. This was an intentional drift using the same current from east to west through the ice, in 1893–1896.

Parry's journey back to London by coach required several changes of horses, and at Durham a delay occurred because the Duke of Wellington had just arrived and commandeered all available animals. Scarcely able to contain his impatience Parry wrote a passionate letter to Isabella in answer to her last letter which had been waiting for him at Edinburgh. It was thirteen pages long giving her his love and his wish to be at her side quickly, especially as her pregnancy was nearly over. He pleaded for news about her condition and the likely date for her confinement; his letter ended with an admission that, 'I am almost wild with joy'.[8]

He reached London on 29 September and went to the Admiralty to make his report; on arrival he was greeted by both John Barrow and the Duke of Clarence, the future William IV, who had been appointed Lord High Admiral of Great Britain during his absence. They received him with 'warmth and cordiality', and 'expressed satisfaction at our proceedings'.[9]

The Duke proposed that he would send him a personal letter expressing his favourable opinion and that Parry should take fourteen days' leave immediately. Before leaving London for Alderley on the next mail coach, Parry sent James Ross orders to ensure that nothing was removed from *Hecla* until after the Duke had inspected her. The Duke's proposed visit being a special endorsement of their achievements, and Viscount Melville was gracious about the result although reminding him, ' I was never very sanguine as to the result of the attempt and I am not surprised, therefore, that the obstacles proved to be of such a nature as even *you* could not overcome'. Barrow's reaction is not recorded but most likely would have been of one of personal disappointment, despite Melville and the Duke's complimentary remarks. Even though Parry had discovered far below the polar icecap an Arctic Sea, it would not become a navigable proposition for the Royal Navy until long into the future when their submarines had such capabilities.

In his letter to James Ross, Parry also described an extraordinary coincidence. As he and Beverley arrived at Somerset House, John Franklin and John Richardson arrived fifteen minutes afterwards. They had been away for three years, travelled more than 1800 miles and mapped most of the coastline of far north-western Canada. After just a few minutes of conversation and mutual congratulation, Parry set off for Alderley in great haste.

A short time afterwards, possibly excusing his haste to leave London, he wrote to Franklin with admiration and in a manner leaving no doubt as to whom he felt had secured his safe return.

> My dear friend. We see your own and your party's confidence in the Almighty. Superiority of moral and religious energy over mere brute strength of body. This journey places you in the rank of travellers above Park (Mungo) and Heavance [sic]. It is a splendid achievement under the blessing of God, of yourself and your brave companions.

He made little comment on his own travels and nothing about his failure to reach the Pole. He ended by hoping that they can both see the 'connection between our discoveries'.

Franklin explained to Barrow and Melville that his intended rendezvous with *Blossom* and Captain Beechey had never happened, although it was later realised that barely 150 miles had separated the two parties at the Icy Cape. It was a disappointment for Barrow for whom the grail of the Northwest Passage remained incomplete – though tantalisingly close. Little more than 350 miles of the northern Canadian coast remained unexplored between Point Turnagain and the bottom of the Prince Regent Inlet. A final but still unknown piece in the complex geographical jigsaw that was the passage.

Parry discovered the entire Stanley family at home except for Isabella, who was travelling back slowly from Anglesey. Their reunion when she finally arrived was, she wrote in one of her journal entries, an 'agony of joy'. She was, Parry happily described to brother Charles, 'in excellent good health'.[10] As soon-to-be new parents they were now absorbed with preparations for their first child and writing to his mother he describes his new daily life as 'seeing and hearing nothing but baby liners, nurses, and cradles', and that he 'has now taken on a new lease of life so well and strong does he feel'.[11]

His special leave was soon over and he returned to London to discover that a new hierarchy had taken over at the Admiralty. Viscount Melville, his long-standing advocate, had been replaced by the Duke of Clarence, who held a level of authority far greater than Melville'. Clarence's wish was to reform and modernise the badly outdated processes and unnecessary bureaucracy prevalent in all parts of the Admiralty. These reforms included proper gunnery training, restrictions on flogging and regular inspections for all naval ships to improve seaworthiness. Younger officers like Parry quickly appreciated many of the changes already enforced and the new more efficient chain of command the Duke had installed. 'The whole concern has been repaired, cleaned and painted, offices and all, and begins to look as a British Admiralty ought to do.'[12] 'Although the Duke is very popular and justly so,' there are, he explained to James Clark Ross, 'exceptions!!!!' He named Thomas Croker as the one of the main obstacles to the Duke's reforms. His own office is at last 'fit for a gentleman and officer to hold.[13] Everyone now minds their own business.' His job as Naval Hydrographer was no longer in any doubt, and after the Duke's visit to inspect *Hecla* before she was paid off, he was given a further month's leave so he could be with Isabella during the last few days of confinement.

Their first son was born in the early hours of 14 November and both agreed he should be called Stanley, since there were already 'too many Edwards in the family and one more would only increase the confusion'. Parry had leave until 1 January at Alderley and enjoyed his free time, the first for ten years. During these weeks he invited James Ross to help him with the journal about their polar expedition and because Isabella especially wished to meet him again.

Their book was intended to be the fourth part of a collection of all the journals and published by John Murray, well before Franklin's next book came out, but disputes arose over the size of the publication. Murray wished to print 'octavo' size, but Parry knew that there was not enough material for this and preferred 'quarto'. Since no artist had been taken with them, Parry asked his brother-in-law, Reverend Edward Stanley, to make a few finished sketches from those he and Ross had attempted. The few finished working drawings were for the engraver, but they provide some indication of the terrible ice conditions and physical exhaustion they had endured with scarcely any of the mood of 'dark and dismal solitude'.[14]

A few days before travelling back down to London he visited Arthur, Edward Stanley's son at his school in Seaforth, staying in Liverpool where the new docks were being constructed as well as the steam railway linking Manchester with Liverpool. Both these large new developments were clear signs of the commercial and technological revolution transforming Britain and which in time would influence his future career with the Royal Navy.

Part II

Chapter 12

A Thankless Office

After returning to London in December 1827 Parry began searching for a house for Isabella and their new family which would be their first home. Caroline Martineau came to help him and by the middle of January he settled upon a small furnished Georgian house at 49, Weymouth Street, at a rent of £300 a year (£25000) with a mews stable large enough for a carriage, although his income was insufficient to keep a horse so the carriage, provided by Sarah, was rarely used. Their small household consisted of a cook, two maids and a manservant.

Isabella followed her husband to London bringing a Welsh girl, the widow of a drowned sailor, who was nurse to Stanley, their new son. Soon after their arrival her brother Edward Stanley and his new wife Henrietta Dillon came to stay. They were also looking for a London house and over the next few months they engaged in a busy social life. Invitations came from Kensington Palace, to the King's levee at St James and dinner with the Dean of Westminster. These occasions were now enjoyable for Parry having Isabella by his side, and while gently amused by Edward's vagueness soon recognised his failings as a husband and brother-in-law.[1]

Parry had reluctantly accepted the position of acting hydrographer three years previously at Lord Melville's request. The department at the time was under the control of Sir John Thomas Croker, first Secretary to the Board of the Admiralty, an antiquarian and an opinionated man who ran it in his own dictatorial fashion for over twenty years.[2] The department had been created during the Napoleonic War, when it was realised the French navy used well prepared and accurate charts. It had never been restructured for any long term peacetime role.

Croker was used to politicians and how day-to-day naval business should be presented to members of the House of Commons but, unfortunately for Parry, he was most reluctant to delegate even routine responsibilities. He was paranoid about reports asked for by the Board and prevented any such documents going directly to anyone, or even through Lord Melville or Barrow. The significance of his department was considerable, being the only source of reliable charts and other essential information for navigators, such as tidal streams and strengths.

It was run very badly because financial pressure on naval budgets by Parliament and an unusual degree of laxity on Viscount Melville's part who allowed Croker to run things in an entirely idiosyncratic manner. This resulted in low morale, an under-resourced and inefficient office procedure and long waits for many new surveys which were taking years to be published. There were too few draughtsmen to arrange and check the final material for the engravers, and insufficient despatchers to get the correct charts to naval ships around the world. This appalling situation was tirelessly defended by Croker on the grounds of expense. Parry's predecessor, Captain Hurd, had even been ordered by Croker to write up notes accompanying the various charts himself to save money instead of using and editing each surveyor's first-hand commentaries; the surveyors reports were paid for separately which was exploited by some who made theirs as long as possible.

Two months before Parry's full appointment Croker issued a departmental instruction that all authority and decision making in future would be held by him alone. He relegated the hydrographer's position to that of 'Director of a Chart Depot for the Admiralty', rather than the initiator, commissioner and final assessor for all maritime surveys.[3]

Parry's appointment in 1823 was only the third up to that time, the first being Alexander Dalrymple, who had combined the Admiralty post with hydrographer for the East India Company. He was succeeded by Thomas Hurd, who had died unexpectedly ten months previously.

The department Parry inherited consisted of only four draughtsmen and one clerk, while the backlog of work included newly completed surveys of the West African coast by W.F. Owen, Australia by P.P. King, and the Mediterranean by R. Smyth, who was simultaneously surveying the entire British Isles for the new Ordnance Survey project. Parry asked for more staff warning that with such limited resources it would take seven or eight years to catch up. But his warnings were ignored and during his absence between 1823 and 1825 the department stagnated even more.

Croker did not disguise his dislike for Parry and he let his superiors and certain newspapers know his feelings. In a letter to the *London Gazette* he wrote:

> I find Parry exceedingly jealous and Barrow no less so. Parry I find the reverse of a manlike open hearted sailor and from a nearer view of his character am convinced that he is as great a *quack* as ever existed. Time I think will make all the Northern Expeditions appear, and not before too long, in their true light of most gross humbug. Parry I believe to be capable of cringing with the most disgusting servility to those above him as well

as deceiving those and trampling on those below him. Thank God our relative situations do not bring us much in contact.[4]

Back in London from his North Pole expedition and leave at Alderley, Parry was delighted to find that the Duke of Clarence had thoroughly investigated the Hydrography Office and discovered, or been told about, his earlier notes and reports, all of which Croker had tried to suppress. On June 7 1827 the Duke wrote in his own hand a Minute which reversed all previous policy:

> Six draughtsmen are to be hired at weekly payments to bring up the arrears of the surveys now at the office and prepare them for engraving. That Lieutenant Becher be employed as suggested in Captain Parry's Memorandum concerning sailing instructions and that an extra clerk be employed under him to assist in the packing and examining of the boxes (for despatch) so that the four permanent draftsmen of the office are not to be diverted from their proper work.

He also asked to see a weekly progress report. This long overdue transformation caused him to write to Isabella praising the:

> immense improvement which had taken place in it [the Hydrography Office] since the Duke's administration of affairs at the Admiralty.[5] Indeed, it is a totally different thing from what it was when I left England, and, as concerns my personal comfort and feelings, improved beyond all conception. It is now in short fit for a gentleman and an officer to hold which was by no means the case when a certain person whose name begins with C was allowed to govern the Admiralty from top to bottom.[6]

Rapid improvements were made in the performance of his department and a new system for updating, withdrawing and replacing old charts in need of correction, were put in place. He decided to improve the separate 'remarks' handbooks attached to each chart in which the surveyors made suggestions for other navigators; he insisted on one standard layout and a simplicity of style and description. He also made sure that 'only surveyors who have a taste and a talent for this kind, should do so. The others should not trouble us with a set of remarks drawn out merely to receive their pay!'[7]

Unfortunately, the intervention of the High Admiral did not last. Intrigues at Court and his scandalous private life eventually forced his father, George IV, with the backing of the Privy Council, to insist on his retirement. The last High

Admiral of Great Britain was removed, never replaced, and went into obscurity in the summer of 1828, until he succeeded to the throne two years later.

Croker once again resumed complete control and made strenuous attempts to minimise those improvements Parry had already completed.

That August, Parry went to France with his nephew Thomas Garnier, and then to Brussels where he was greeted with both public and government acclaim, awarded several decorations, and made an Honorary Member of the elite Royal Institute of France. He was also invited by the Duke of Orleans to his family home in Neuilly, where he met and played with the princes and princesses on a boating lake. The children insisting on playing a game in which the lake was the Arctic Sea full of icebergs which, with his help, they had to avoid.

During the voyage back from Rotterdam the ship met a severe storm and, entirely due to his navigational experience and seamanship Parry managed to avoid shipwreck upon the Goodwin Sands. This trial was soon followed by two more. The tragic death of his infant son Stanley, only 1 year old, and then the death of a prematurely born second child, after a complicated confinement for Isabella. These tragedies though seem only to have strengthened his religious convictions.

He pronounced their deaths as an act of God, over which, he wrote to Lady Maria, 'it brought us much, very much, nearer to God through Christ'. She swiftly wrote back, 'you are doing her [Isabella] an injury by this adoration'. Eventually these differences quietened down, although in a later letter he wrote, 'what should we say, if God as easily entered into judgement with *us*, as we do with one another? Alas! What is left for us in this, and all other such cases, but to forgive one another and pray that he may forgive us also?' It was one inevitable consequence of Lady Stanley's innate eighteenth-century scepticism meeting Parry's nineteenth-century evangelism.[7]

Parry returned to the Stanley house at Penrhos near Holyhead for a brief stay before travelling back to London at the end of September 1828, where he found Croker to be on leave and that, 'everything was in a remarkably quiescent state as if nothing had happened, just as people live under a volcano just after an eruption. In fact the old regime seems to have succeeded so quietly and smoothly that the change is no longer a matter of conversation or remark.'

Melville was also away in Scotland but Parry has made up his mind to 'explain my views to him' when he returned. This was a turning point in his career, but a decision Parry knew he needed to take; he decided to resign but, before he could speak to Viscount Melville over the reinstatement of Croker, the Admiralty Board decided that officers who held salaried appointments such as Parry would no longer receive half pay, a traditional way of retaining officers even when commissions were unavailable. He wrote to his brother Charles, 'I am losing both health and money (about £200 per annum) in my present situation. There is no other situation to get or to give; witness Franklin still unemployed.' In a letter to Lady Maria he mentions the proposed private expedition being planned by Captain John Ross at a cost of £20,000 (£1.4 million today), but underlines in his letter the words, 'do not be alarmed – I do not propose to take any part in this Affair'.[8] He did not like the design of the boiler and engine combination which was he felt too radical, nor the paddle wheels which were too weak, and decried the lavish manner in which the expedition was being fitted out. Ross had raised money from a private sponsor, William Booth the gin distiller, and wanted to commission his steam-powered paddle steamer the *Victory* for both her speed and her capability, on paper at least, to break through ice. Parry, after a single visit to *Victory* fitting-out on the Thames, felt that the innovation had potential, but that the paddle wheels which could be lifted clear of the ice if necessary, and several new pieces of equipment such as her boilers, were insufficiently tested and were, 'just novelties which should not be tried for the first time on a service such as this'. His advice was sound and the engine was abandoned on land only weeks after they had crossed the Atlantic.

He was also perceptive enough to propose that if their initial sixty days' supply of coal was insufficient to get through Regent Inlet, they could stop to collect the coal he had abandoned on *Fury* beach.

The Ross expedition was ultimately saved by the fuel, food stores, several small boats left behind, and timber from *Fury*'s hull. Ross and his crew abandoned *Victory* after being trapped by thick ice at the southern end of Regent Inlet for three freezing winters and used these abandoned stores to build a weatherproof cabin heated by coal for one final winter before being miraculously rescued.

Soon after visiting *Victory* in April 1829, Parry received a letter from the Australian Agricultural Company inviting him to become their Commissioner in New South Wales. This invitation had been made earlier to John Franklin, although whether each knew of the other's involvement is uncertain. Both decided initially to refuse, but after a short period of reflection Parry changed his mind. This had partly come about due to advice from Barrow and Melville. The Colonial Office and the Admiralty were becoming concerned about the economic viability and security of the new British Colony in New South Wales.[9] Several expensive and ambitious efforts had been made to establish another new colony in the Northern Territory at Darwin as a settlement and naval base resembling that at Singapore. These had all failed and British control over Australia was considered vulnerable to invasion by a foreign power, especially France. A resourceful, reliable, competent and purposeful individual was urgently needed to bring the precarious financial situation under control.

It was an unusual opportunity, but it offered Parry a clear way of increasing his income and to remove some long-standing debts.[10]

He sensibly asked brother Charles's opinion, who in his turn recommended their old family friend, a solicitor called Barenshaw, for advice. Parry took his suggestion and asked Barenshaw to act as his intermediary.

The outcome was a substantial salary of £2,000 a year (£136,000 today), a house, a clerk or secretary, two female servants, £800 (£54,000) in contribution towards the cost of the passage for his family both ways, and a 'fitting out' payment of £200 (£14,000) for clothing and various household essentials. After a long negotiation the Company also agreed to provide an annuity of £300 (£20,000) per annum for life on his return and £100 (£7,000) for Isabella if he died in their service. He now only needed to learn, before finally accepting, whether he would remain on his captain's half pay and the list for suitable commissions when he came back to England. He still considered himself primarily a naval officer, whatever else he needed to do to earn income. This application could only be authorised by Melville and Barrow, and by a lucky chance he received a letter from Melville in mid-April informing him that he had been awarded a knighthood along with John Franklin. Parry was given what he wanted, including leave of absence for four years with his half pay. After an interview with Melville he stayed on for dinner to discuss his replacement at the Hydrographers Office. He strongly recommended the Irishman Captain Francis Beaufort KCB FRS (1774–1857), who became the best known and arguably the most successful hydrographer the Royal Navy ever appointed leaving over 2000 completed charts for his successor. Parry's grandson, Sir John Franklin Parry KCB FRGS (1863 - 1926) also became head of the Hydrographic Office between 1914 and 1919.

One serious difficulty required careful management. Lady Stanley's reconciliation to these new plans and Isabella's absence for four years. She, unsurprisingly, was horrified at her daughter being 'transported to Botany Bay', as she described it, but fortuitously, the Stanley family connections included Bishop Turner of Calcutta, in whose jurisdiction New South Wales then lay, and both Edward Stanley, Sir John's heir, and Turner were strong advocates for the new colony and won her round. His award of a knighthood may have helped placate her feelings since she now recognised that his social position, though not equal to hers had risen, and the news that his salary would include a special annuity for Isabella should anything happen to him, were sound arguments in favour of his case.

Isabella's reaction to the new situation, which was entirely different from that role she expected as the wife of a successful naval officer, proved her absolute devotion to her husband even though it meant uprooting from her friends and family for a long period. Exchanges between Isabella and her sisters at the time of her engagement to Parry imply that she had been anxious to gain greater freedom from her overbearing mother for sometime, and now, after the death of her first two children, the move abroad allowed them both to leave behind painful troubles and sad memories. Her doctor's advice was supportive and he endorsed the decision on the grounds that the many risks from virulent diseases such as typhoid, cholera, scarlet fever and tuberculosis prevalent in London at the time must be avoided to strengthen her own indifferent constitution, as well as those of the small children. The warmer and cleaner environment of Australia would be most beneficial.

At the Court levee on 29 April 1829 at St James House, Parry and Franklin, in full dress uniform, were knighted by King George IV. Parry was 39 and Franklin 43. Their joint citation read:

> the King was pleased to confer the honour of Knighthood upon Captain John Franklin of the Royal Navy and late Commander of the Northern Land Expedition and Captain William Edward Parry of the Royal Navy late Commander of the Expedition for the Discovery of a North West Passage.

After the ceremony they drove to 76 Baker Street, where Sarah and his two unmarried sisters had finally moved in 1828, as he had wished years earlier. They were announced as Sir Edward and Lady Parry; Sarah now 80 years of age, wept for joy.

Franklin continued petitioning both Barrow and Melville for another Arctic commission, this time to close the remaining portion between his two earlier land surveys and explore eastwards from Point Turnagain. Some members of the Admiralty Board, and most of Parliament, however, had lost their initial enthusiasm for new expeditions. Franklin's proposal was rejected within twenty-four hours of being submitted.

Franklin had married Jane Griffin in 1828 (See Fig. 15), who had been a close friend of his first wife, Eleanor, and this new marriage was highly fortuitous since she came with £7,000 of her own and a settlement from her father of another £3,000. The combined equivalent of £680,000 today. She was an intrepid and inveterate traveller, having ventured on her own into the near East and other parts of the Eastern Mediterranean, and faced hardship and danger in remote places with unwavering self-confidence. Her husband's fortitude and stamina during his long and physically demanding journeys, was matched by her own.

On 1 July, Franklin and Parry, along with their wives, met for the same award of Honorary Doctorates from Oxford University, an event which gave both men great pleasure. The poem written to honour these awards was entitled,

'Voyages of Discovery to the Polar Regions'. The same honour was bestowed on that day to John Eardley-Wilmot, Parry's brother-in-law, a lawyer, politician and strong Whig supporter and he had now remarried. He would eventually replace Franklin in 1843 as Lieutenant Governor of Van Diemen's land, a penal colony now known as Tasmania.

In the first of two letters sent to Charles in early June, Parry is relieved that the first payment of his new AAC salary will 'see him out of debt. I therefore start fair once more.' He explains that Isabella is, 'thank God pretty well. She is in the family way and we may expect a "young convict" in less than two months after our arrival in Sydney.'[11]

In his second letter he wants them to meet at Summerhill and asks that he will become his attorney and that his brother-in-law Joseph Martineau will be his agent over money matters while he is abroad. He was then working at the AAC office in Kings Arm Yard reading papers, while Isabella was shopping with Caroline Martineau, 'trying to foresee all the things they might need for the antipodes. We have decided to take Fido the dog with us.'

They were due to depart Gravesend on 15 July on the *William* from Gravesend, a voyage expected to take at least ten weeks. Travelling with them would be a new cook Sarah, James, a manservant, a new lady's maid, and Parry's new private secretary, Henry Darch. Also on board were William Burnett, the newly appointed agricultural officer for the Company with all his family. Some important letters from Charles clearing up financial arrangements between them would be addressed to Devonport, which they would visit before finally sailing to Australia.

To his great satisfaction Parry was told that his successor at the Hydrographic Office would indeed be Francis Beaufort, a widely respected and trusted officer who he had recommended to Melville. Beaufort would be set free of Croker under several new reforms instigated by Sir James Graham, the first Sea Lord. He would have complete autonomy for a fully staffed department, separate entirely from outside interference and mandated to prepare and manage his departmental budget and schedules of work – all principles and reforms which Parry had fought for. The *William* arrived in Plymouth, which Parry had last visited twenty-six years earlier as a very junior officer in the Channel Fleet. They discovered John Barrow, Lord Melville and Lord Yarborough also paying a visit and walked with them along the newly constructed Plymouth breakwater, later dining on Lord Yarborough's magnificent yacht. They sailed for Sydney on 31 July 1829.

Chapter 13

The Commissioner

The Australian Agricultural Company had been established under Royal Charter in 1824. Its purpose being the 'immediate introduction of capital, and of agricultural skill, as well as the ultimate benefit of the increase of fine wool as a valuable commodity for export ... for the further advancement of the Colony.'[1] The Earl of Bathurst, Secretary of State for the Colonies, sent the decision to Governor Thomas MacDougall Brisbane (1773–1860) stating that the AAC directors would now despatch Robert Dawson as their principal agent and surveyor after purchasing Merino sheep from the Continent. He would travel with the sheep and a small group of capable men to New South Wales to survey and select land suitable for the purpose. The extent of the grant area given by the Government to the Company was one million acres (See Map 4). It stretched from Port Stephens and Newcastle on the coast up to the Liverpool Range in the north.

The directors had already appointed a local committee to supervise Dawson but decided to hire John Macarthur as their chief adviser, who, although useful due to his sheep farming background, was a self-interested and very volatile individual when either giving or seeking advice. Dawson had been persuaded to recommend that the new Grant should be centered on Port Stephens, a wide inlet approximately 100 miles north of Sydney and was actively discouraged from surveying other much more suitable parcels of land beyond the Manning river, and west from the confluence of the Gloucester and Barrington rivers. Beyond the Liverpool hills was an exceptionally fertile area of land, the Liverpool Plains, while further west lay the fine pasture land of the Hunter River basin above Newcastle.

Very soon it became clear that the land Dawson had chosen was unfavourable for sheep. The new animals bought both from the Continent and in England were dying in large numbers and a three-year-long drought had worsened the situation. Dawson was dismissed and the board of directors then decided to appoint a Commissioner who would take charge of all their interests, resurvey suitable alternatives for grazing and negotiate with both local landholders and the British Government for an exchange of suitable land. Following this difficult transaction he would build up a sustainable and profitable sheep breeding and wool exporting business from the 30,000 sheep that were already there. This all to be accomplished within his contract period of four years.

The task was a large one, not helped by the deep animosity held by many of the smaller landowners and farmers towards the AAC. The *Sydney Gazette* declaring the Company would 'one day or other pour forth its cattle from a thousand hills, by means of which every minor stockholder will be swept away in the inundation'.[2]

The London directors failed to remove or replace the local committee and, rather worse, to state unequivocally that their new Commissioner had complete autonomy. These weaknesses in his position created acrimony, procrastination and steadfast refusal by several committee members to take instructions and so the financial losses continued. Most of the committee members had been selected by Macarthur and were closely connected in business or related to one another. Many were deliberately obstructive to Parry's ideas and reforms on principle. Ruthven's 1826 List of Proprietors shows six Maccarthurs on the list of the New South Wales Shareholders. Parry's appointment was due to his reputation for honesty, determination, surveying knowledge and an understanding of sheep and sheep breeding. He had had contact with several large landowners in Britain, of whom one was Lord Bathurst, the current Colonial Secretary. Parry, he knew, could be trusted to follow instructions completely but was capable of taking important decisions on his own initiative when required. Letters to or from London with any reports and suggestions took six months to travel each way; his integrity and decisiveness were therefore essential qualities and his naval training had two other important benefits. He kept meticulous daily records of every action and decision he took and why, and was used to dealing with men who needed a firm hand and gave them an exemplary example of probity and honesty.[3]

The problems he faced on arrival were the acute shortage of working capital caused by Dawson's weak financial controls, the excessive number of indentured servants and officials, and the shareholders' failure to honour their financial commitments. The indentured servants, such as clerks, had mostly been employed on long-term contracts, some of which still had years left to run. The proprietors, which included all of the local Committee, had never paid for their shares.

When the Parry family arrived on 23 December 1829, they were invited to Government House to meet Sir Ralph Darling (1772–1858) the Governor General. They both made a highly favourable impression on the Governor and his wife who wrote after their first meeting: 'he is a most excellent person altogether and his manners have so much the heartiness of the English sailor that he and our children are great friends already'. Staying at Cummings Hotel in Macquarie Place, they were visited by most of the local farm proprietors as well as State officials, including Major Mitchell the Surveyor General, the Chief Justice Sir Francis Forbes, Archdeacon Broughton, and the old and cantankerous John Macarthur, chairman of the local AAC Committee even though it had been reported that he was too ill to travel.

It transpired that he was yet another old acquaintance of Parry's father Caleb, and was entirely happy to give Parry his plangent opinions about the Company's failures and shortcomings, the main criticism being that the AAC stock must not be cross-bred but kept pure. He knew that breeding 'in and in' as he described it, was of no importance providing the ram was good.[4]

As well as Dawson, whom he utterly despised, Macarthur wasted no time in telling Parry that Mr Macleay, Clerk to the local committee, acted consistently against the Company's interests and that the 'entourage' of indentured and assigned servants was far too numerous for their purpose. He also felt that Dawson had placed the main offices of the employees in the wrong place, 'in a swamp and opposite a mud flat in the sea – so that it must catch the drip between the two hills and therefore could not be dry during the rains!' This was at Carrington, and Parry found his views to be entirely accurate.

Macarthur and Darling supported the directors' plan to exchange parts of the current grant territory for better holdings elsewhere. Darling sent a letter of confirmation to Lord Bathurst of his approval and also asked for confirmation that a proposed transfer of the coal works at Newcastle to the AAC might be ratified, allowing Parry to create some badly needed extra revenue for the business. The works were quite primitive, with coal being 'cropped out' from a high cliff face. Three seams were being worked but only one produced coal of any acceptable quality. There were thirty-one employees, all convicts. Transport was by bullock carts to barges moored out in the river.

Parry then called on Major Mitchell, the Surveyor General, to study his maps and to ask him whether he should look at territory north of the Manning River, and to the west of the existing holding, held under the jurisdiction of the Corporation of School and Clergy. Mitchell expressed no objection but warned him that earlier settlers had already sought permission to purchase the former area while counter proposing that he considered land at Moreton Bay which could be suitable for sugar, or even tobacco growing.

Isabella was undergoing another difficult pregnancy and still recuperating from the long sea voyage from England. They decided that she should remain at Government House while he set sail immediately for Port Stephens on the *Lambton*, the Company cutter, with Mr Darch and two servants. A dog cart carrying some of their possessions went with them, and Parry would begin inspection of the existing grant land without delay. The day before his departure on 5 January 1830 an announcement was placed in the *Sydney Gazette* that unless long overdue payments from proprietors of the AAC were paid quickly, their shares would be forfeit.[5]

A strong southerly running current delayed their journey and they reached Port Stephens at midnight on 8 January. The next day he wrote to Isabella describing the wild beauty of the place, with a forested wilderness with cedar trees but with less rock than at Sydney and the immense expanse of the wide estuary above Soldiers Point, splendid at high tide but, as he quickly pointed out, 'a gay deception, since almost the whole was a mud flat a large proportion of which is dry at low water on spring tides'. Port Jackson was deeper and much healthier (See Fig. 16).

They were transferred into a smaller boat to reach Carrington, or Carrabean in the aboriginal tongue, where he was met by James Ebsworth, one of the younger clerks, and shown his cottage at Tahlee, just outside the township; this had been 'carefully railed round and the land smoothed and levelled beside it'. Parry walked about visiting every part of the settlement, meeting the inhabitants visiting the stores, the neglected corn mill, the houses, the barracks and even the jail. All were noticeably in various stages of neglect and disrepair. He then went to the hospital which he described as 'a most wretched shed', full to capacity with beds for a mere six patients.

There was no school, but after conducting divine service in the church the following day Parry asked the small congregation for their views and they all agreed that one must be started.[6] This decision to build an educational foundation for the settlement is spelt out by his journal entry: 'I expect no success in our undertaking, unless an improvement first commence in the religious and moral aspect of our community. It is an absolute moral wilderness at present.'

Just as he had persistently worked to give a basic level of education to his naval crews as well as the understanding of Christian values, he was certain that this step was a priority, notwithstanding all the many difficult and serious problems he faced. He reckoned that education would be the most effective method of building a more responsible, hardworking, contented and law-abiding community for the future.

A shipment of Company wool was soon to be loaded for Sydney on the *Lambton*, but bad weather and the unsuitability of the wharfage arrangements at Port Stephens showed him yet more of the underlying problems undermining the commercial fortunes of the Company. The wharf at Port Stephens had been built in a place where, apart from at high water, there was too little depth for vessels to load or unload.

For the next week he and Ebsworth toured the territory on horseback, with Parry suffering from a bad attack of dysentery. When they returned to Carrington the *Lambton* arrived at Port Stephens carrying the news that Isabella had given birth prematurely to twins. Parry went quickly on board, wool bales were rapidly loaded and *Lambton* sailed back to Sydney, leaving Ebsworth in charge. During the voyage he wrote to his mother telling her the news:

> Isabella was on the 14th safely brought to bed of twins – a boy and a girl – and they are all doing well! They were born about three in the afternoon the girl ten minutes before the boy. The boy is smallest and has required great care – hot bottles etc. etc –and indeed we owe its life under Providence, to Mrs Darling suckling him herself for two days and nights, tho' herself in bad health. I cannot express to you the affectionate attentions we have received from these dear people.[7]

He did not add that Mrs Darling had had both children baptised as Edward and Isabella very rapidly, doubting that they would both live. But the children survived and began to thrive and it was Isabella who raised more concern by her weak condition while he was laid low by exhaustion and stomach complaints which lingered on so that he felt 'improved but not yet stopped'. A treatment and a consequence he blamed on treatment for his dysentery by the Company doctor at Carrington.

During their joint convalescence at Government House Parry continued questioning Governor Darling about each of the Company's officers, including Mr Bunn the Company agent in Sydney, the shareholders, and of John Macarthur, who visited him again on this new occasion to persuade him to move the Company headquarters from Carrington to Sydney. After seeing the poor conditions at Carrington and Port Stephens Parry might have been tempted, but he was now more sure that with improvements to its buildings and roads, and a proper dock at Port Stephens, Carrington – being close to the grant lands – was a much better base from which to inspect and manage the superintendents of the widely dispersed farms.[8]

A month after returning to Sydney he went back again to Port Stephens leaving Isabella and the new twins at Government House. He returned just as the final wool shipment for the season was being sent away and there was an outbreak of sickness among the employees. The weather was very bad with incessant rain proving the unsuitability of Carrington as headquarters of the Company due to flooding around the settlement and its hinterland. The main roads were in a terrible state. Undaunted and travelling continuously, he finished his interrupted tour with Ebsworth, and organised work at their cottage at Tahlee ready for Isabella and the family. He also organised and re-equipped the group of convict constables, who were costing the Company £500 a year. He also dismissed the Watch House keeper for laziness and demanded from the Archdeacon the appointment of a Minister for the settlement. Efforts to bring more people to church gradually began bearing fruit and the congregation was, he felt, now more 'respectable', and included women and their children. Then he studied alternative places for a new cargo wharf for loading the wool cargoes onto *Lambton*. The wharf that Dawson had built being useable for only a few hours at high water.

His opinion of the colony at this time was outspokenly expressed in a letter written to Lady Maria:

> everything, good and bad, in and about this country is grossly exaggerated. Half that has been published is a wicked and gratuitous lie. 'Sir,' said an old settler to me the other day, 'we are a colony of liars.' 'Sir, I believe so,' I *might* have said – but only thought so... It is certainly a country in which one must always be on the defensive – I do not mean as to the protection of life and property, respecting which there is no particular fear – but in the common business transactions in which I am daily concerned.

To a man such as Parry, with deeply held views about truth, integrity and punctuality, the attitudes and behaviour of the indentured servants and officers of the Company were incredible. In the Royal Navy an order always had to be carried through to the letter and without challenge which he soon found to be entirely at odds with most of his employees. As the first AAC Commissioner he had anticipated initial difficulties with understanding and compliance to his orders but the responses he got were exceptionally grudging, defiant, or just plain negative from the majority of them – both junior and senior.

The land he saw that impressed him most during his tour with Ebsworth was around Barrington Flats. It had good alluvial soil, plenty of clean grass, magnificent ferns and many wild apple trees. Parry was delighted by this new vision of the potential as well as the beauty of the country and of the many agricultural possibilities it offered. The extent of the original company grant stopped at the Barrington River, just short of the really good land on the opposite bank. This boundary, Parry noted 'had been shamefully fixed'.[9]

On the last day of March 1830 Parry went down to Government House and brought Isabella and the children up to Tahlee, 'This quiet and lovely place – our future residence.' By now there were 500 people living in and around Carrington, half of whom were convicts. There were also 1,000 aboriginal natives living behind the settlement in their own camp. It was not, Parry soon decided, the convicts that were of concern but the free men who lacked the pioneering spirit and he wrote to his mother again, giving her details of the house and the children: 'everybody (almost) has been minding their business a great deal more than the

Fig 1. Sarah and Caleb Hillier Parry. c.1778. Mezzotint. (*Artist unknown*)

Fig 2. Summerhill. c.1858. Photograph. (*Artist unknown*)

Fig 3. Wreck of the '*Magnificent*' off the Boufolc Rock Brest. 1804. (*Engraving. J Schekty*)

Fig 4. Portrait of Sir John Barrow. c.1834. (*Oil on Canvas G.T. Payne*)

Fig 5. Portrait of Lieutenant William Edward Parry. 1819. (*Oil on Canvas. Sir William Beechey RA*)

Fig 6. Portrait of Sir Joseph Banks. KB FRS. cf 1803. Oil on Canvas. (*William Daniell*)

Fig 7. Portrait of Sir John Ross. 1829. Oil on Canvas. (*J. Green*)

Fig 8. Portrait of Captain Sir John Franklin. c.1830. Oil on Canvas. (*William Derby*)

Fig 9. Ross and Parry introduce themselves to Inuit at Elah. 1818. (*Coloured Sketch. John Sacheuse*)

Fig 10. HM Ships *Hecla* and *Griper* blocked up by ice in Winter Harbour. 1820. (*Engraving. William Westall*)

Fig 11. Prismatic Cross with Full Moon. cf 1854. Engraving. (*Charles McDougall*)

Isabella.

Fig 12. Isabella Stanley with their two eldest sons, Edward and Charles. Date unknown. (*Engraving and photograph. George Richmond*)

Fig 13. Canoe of the Savage Islands. 1832. Engraving. (*Captain G.F. Lyon RN*)

Fig 14. The Wreck of the *Fury*. 1824. Engraving. (*Henry Head RN*)

Fig 15. Jane, Lady Franklin. Age 24. Date Unknown. (*Engraving. A. Romilly*)

Fig 16. Port Stephens. NSW. 1826. (*Coloured Engraving. James Wallis and Joseph Lycett*)

Fig 17. Newcastle NSW. 1828. Engraving. (*Artist Unknown*)

Fig 18. Catherine Hoare with her two daughters, Louisa and Priscilla. 1839. Watercolour. (*George Richmond*)

Fig 19. Portrait of Sir William Edward Parry FRS. 1842. Engraving. (*H. Adlard*)

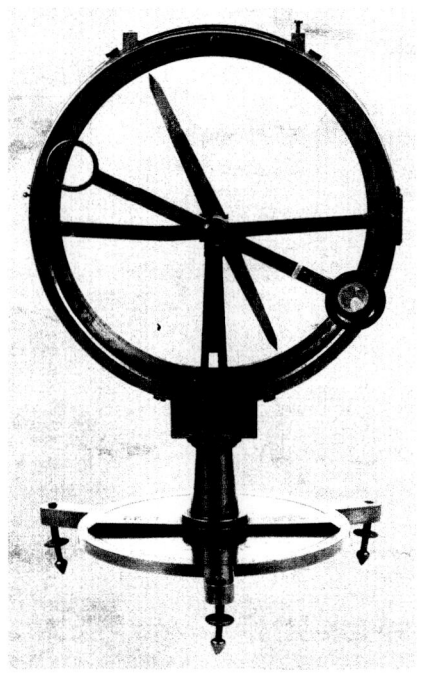

Fig 20. Dip circle or dipping needle. c.1840. Photograph. (*Robinson*)

Company's'. In his journal he wrote: 'I begin to be very sick of the free servants of the Company and I verily believe that the work would be nine times out of ten be done better by convicts.'[10] Unfortunately, even though more convicts were arriving on every ship, there was less chance of these being assigned to the AAC, as under the existing laws new settlers were given first choice. The company's development was seriously hindered by this shortage of manpower, as well as the money it had to contribute to the running costs of the police constables. Throughout his posting Parry argued with the British government about both these important issues. His attitude to the convicts themselves by no means indulgent, was reasonable and just, and he expressed his views at the Royal Commission on Transportation of 1838: 'I think a man should never be placed in a situation where hope is entirely taken away from him.'

His confrontations with management in the Colony were matched in outspokenness by his letters to the directors in London, especially over the two appointments on which he had not been consulted. These were the new surveyor hired in addition to Armstrong, and a mining engineer called Henderson. The reason for the latter appointment being a new viability appraisal for the Newcastle coal mine and was his second visit to the colony. It followed on from an agreement by the British Government that the Company could have a thirty-year monopoly and a further grant of territory if they developed the mine and its revenues. The directors hoped, in spite of Henderson's inconclusive initial tests, that this might yet provide a way out of their severe financial difficulties. He arrived with new and expensive equipment, and after meeting Parry at Newcastle he began to drill for new seams of coal lower down the cliff. Parry meanwhile, anxious to find the land to exchange for the poor territory chosen by Dawson, decided on an expedition to the Hunter River, visiting on the way successful agricultural businesses growing tobacco, maize and wheat near to the settlement at Stroud. Here he met individuals such as Mr Jones, the overseer of one rich parcel of land and, as before at Barrington, his opinion of the country and its true potential increased, along with his conviction that compared with the likes of Mr Jones, the majority of the Company officers were lazy and useless.

Parry acted decisively by insisting that Mr Barton, the resident accountant and worst offender in Carrington, would in future report directly to him and not to the Committee.[11] He followed this by forcing Dr Stacy, the incompetent practitioner who had mis-prescribed his recent stomach sickness, to report to Dr Nisbet, recently arrived from Sydney.

Slowly he stripped out the worst abuses by officers of the Company, insisting on adoption of proper accounting procedures and a rigorous inspection of the accounts for each part of the business. Nothing, however minor, escaped his notice and for the first time in its existence a properly conducted valuation of the Company's assets, including stock and buildings, was sent to London.

Isabella settled into the cottage at Tahlee which she managed to turn into a comfortable home. With her husband away frequently she furnished the simple cottage, which at first seemed strange without carpets or the usual plush furnishing they had in England, and began to enjoy their small garden which, with help from her mostly male convict servants, she transformed into a place of peace and contentment. She wrote to her sister Lucy that,

> I never wish to have nicer nor better servants. It is quite a new thing for them to meet anything like kindness or encouragement and I feel sure that by what I have already seen that far, far more may be done by kindness than by harshness. I am making acquaintance by degrees with those who work near us, and shall indeed rejoice, if we are permitted to be the means of good to them.

When Parry was back home they started a new school at their cottage for convicts who wished to learn to read and write, and formed a choir which practised every week at Tahlee and sang every Sunday in Church, accompanied by flutes.[12]

Since the Company employed two surveyors, Henry Dangar and Charles Hall, Parry sent them out to survey different tracts of land suitable for an exchange. He concentrated on the administration of the Company, inspecting all the books, and writing detailed reports for London. While preferring to be on horseback with them rather than desk-bound in the dining room at Tahlee, he trusted Dangar's ability and good judgement so that when he came back to report that the land to the north of Taree (see Map 4.) and the Manning River was unsuitable, but that if he should go west and survey the area between the existing grant and the land to the north of the Great Dividing Range, Parry was confident he would find the answer he needed. Charles Hall and Dr Nisbet meanwhile went to the Avon creek boundary within the existing grant territory after the lambing season ended.

News now arrived from Newcastle that August that a 5ft thick coal seam of good extractable quality had been discovered, so on 2 September Parry went to see for himself. He saw that a number of basic improvements would be required, including new housing for extra convict labour to work the mine, a proper school and improvements to the police barracks. A couple of powerful steam engines were ordered for drilling and then building a mine shaft; they arrived on the *Nereus* a few weeks later. The site of the new engine house and a bigger wharf were his next two priorities and such was the urgency felt by Parry and Henderson that they set about planning and construction without getting approval from London. This action proved both timely and rewarding as the British Government had recently decided to withdraw its control over selling the coal and all profits would now go directly to the Company.

The new buildings, wharf and mineshaft were rapidly finished in December and an initial cargo of coal was loaded onto the *Sophia Jane* steamboat, the first ship of its kind to trade along the coast of New South Wales.

Christmas 1830 was the first anniversary of their arrival in New South Wales and special celebrations were organised. The settlement at Carrington, badly run down when they first arrived, now had better housing, enlarged warehouses for wool storage and any other agricultural produce, decent airy barracks for the military detachment and a new school house. A specially made tent was erected on the shoreline flats, decorated with flowers and flags for the prize-giving ceremonies and a special dinner given for parents of the schoolchildren and the officers of the Company with their wives. Fifty attended this event on 23 December, which Parry described in his journal as 'putting a little English feeling into them which was in every way desirable'.[13]

After the dinner games were played and races run, followed by a big tea party for the children only cut short by signs of an approaching hurricane. The next day, Christmas Eve, Isabella gave Christmas boxes to each of the indentured servants and at midnight the new choir went round all the houses singing carols.

The convicts were all given extra allowances, including half a pint of rum dispensed to each at different times of day and to separately avoid, as Parry put it, 'any possible noise or disorder'. The convicts were also given a day of rest, their only one for the entire year, and invited to play an improvised cricket match followed by buckets of tea. His own sympathy for the conditions of these people is made clear in a journal entry for 29 December, 'these unfortunate men enjoyed it very much, and are, I really believe, grateful for this enjoyment which can be NO loss to the Company'.

His determination and constant activity meant that even before New Year he had ridden back to Newcastle to inspect Henderson's mining works. Much had been accomplished in a short time and the simple but effective engineering of the gravity-run railway leading from the new mine down to a much improved new wharf, satisfied him considerably. He returned to Tahlee to be with his family for New Year's Eve, although the following morning he was forced to adjudicate in more disputes, suffering more insolence from clerks and receiving letters from Barton and Burnett full of contemptuous and ill-mannered comments. For a full twelve hours behind his desk he answered their unhelpful responses to his earlier instructions and demanded from each more precise information. James Ebsworth, the man who had first greeted him at Carrington, stayed at work alongside him offering 'valuable and cordial assistance', although he was soon to return to England in the near future.

These angry confrontations and the long hours spent contesting his decisions were affecting his health, giving him severe back pain which he described simply as lumbago, and sharp periodic pains in his abdomen. Any travel on horseback over long distances aggravated these problems. Isabella, whose health was now improved considerably, was pregnant once again and they agreed she would go down with a nurse and servants to Sydney where there were better medical facilities. Needing a house to rent, Parry set sail on the *Lambton* following the year's final shipment of wool from Port Stephens.

He planned to spend considerable time in Sydney with the Surveyor General looking closely at his maps of land around Liverpool and Goulburn Plains north of the Great Dividing Range, and the flat land between these and the Company's existing grant (See Map 5).

Chapter 14

The Great Dividing Range

In 1831 Captain Sir John Franklin was on Corfu having been sent there by the Admiralty to command HMS *Rainbow*. As noted in chapter 12, he had married Jane Griffin in 1828; the capital from her dowry provided enough for them to travel and for Franklin to wait for suitable commissions to be offered by the Admiralty. They visited Russia and many other European capitals during this time. Meantime, the Royal Navy went on reducing the numbers of its ships, officers and men. The dominance of the British navy was so great that no other country was capable of challenging its supremacy and Parliament was disinclined to continue passing large budgets unless it was absolutely essential or unavoidable.

Eventually Franklin was offered an unusual peacekeeping mission. Based in the Eastern Mediterranean he was to provide diplomatic support and encouragement to the newly created Greek State, and to protect the mercantile interests of its British residents. The role combined not only command of a naval force to deter Russian and French naval forces being used to exploit factional interests and extend their influence with the new Greek government, but also to keep the peace and prevent local warlords from seizing control of ports and trading routes. He succeeded in steering a way through these complex diplomatic and potentially dangerous situations by his, 'dogged British pertinacity' and international fame and reputation.[1]

Parry showed many of the same qualities and his job as Commissioner was constantly obstructed by the local Committee shareholders, and most of the indentured officials he had inherited on appointment. These individuals were typically self-seeking, frequently incompetent, prone to drunkenness and very lazy. His efforts to remove, or at least minimise, the damage caused in the Company by these men continued, and he then learnt that a replacement for his most important ally, Governor General Ralph Darling, was arriving soon from England.

During the summer of 1831 and before Isabella was about to give birth, Darling had announced he would be returning to England. He had been victimised in a hostile personal campaign headed by a large landowner called Wentworth, who was leading a party aiming to create a new Republican Party.[2]

Darling governed with a stern hand and had never tried to conciliate with or allow any radical political activity. He had frequently called them an undemocratic rabble and his unpopularity among this vociferous and dangerously subversive group had steadily increased. His recall had been solicited by other supporters for months through British newspapers and in diplomatic circles within Whitehall. His replacement was an Irish soldier called General Richard Bourke (1777–1855) whom Parry came to esteem highly and described as, 'no ordinary man'.[3]

The Darlings left Australia before Bourke had arrived and Parry felt their loss. They had shown the Parrys kindness and hospitality, which had been so essential during Isabella's first confinement and then convalescence immediately after their arrival. Fortunately Isabella gave birth without difficulty to a daughter they named Lucy, chosen in memory of Lucy Stanley, Isabella's favourite sister. His letter announcing this decision described her namesake as a 'little dark haired and dark eyed thing'. He concluded his short letter to the Stanleys, dated 20 September 1831, saying he was 'off to Port Stephens again and could not add a line more'. There was no mention at all of Isabella's health!

A letter sent soon afterwards to his mother took a long time to reach her and Sarah had died, at the age of 82, before it was delivered. Sister Caroline described the last moments of her life in her letter accompanying a box of mementos; Parry wrote immediately after reading her sad account, in July 1832, to his brother Charles. It was an emotional letter describing their mother as, 'dear, tender, affectionate and pious'. His Christianity was again intensified by the loss of his mother a very great distance away, whom he describes as 'sainted'. In this letter he proposes that their two unmarried sisters, Matilda and Gertrude, are provided with funds supplied by each of them to pay for servants and to 'secure their comforts'.[4]

He next received a very welcome letter from the London directors confirming that James Ebsworth would return to the colony as his deputy and chief accountant. Mr Barton was to be given 'six months' notice to quit the Company's services'.[5] He dryly observed, 'I am truly thankful the directors have in this last arrangement anticipated me and I them.' He had in fact already given the tiresome and unhelpful Barton notice.

Returning again to Carrington he found to his satisfaction, 'everything was going on with energy, good order and apparent good humour' in the sheep shearing and that he had 'not a single petition to deal with'. The country though was parched, and there were rumours of large bush fires up country. On his way home he quickly visited Newcastle to attend the official opening of the new wharf. The Company could now expand the production and sale of coal, employing the first steamboat starting to trade cargo along that coast and to trial some shipments to Van Diemen's Land (Tasmania).

The company surveyor, Mr Dangar, had recently returned with encouraging news. There were two tracts of land beyond the Liverpool (or Great Dividing) Range, one of about 200,000 acres on the Liverpool plains, the other similar in area lay along the banks of the Peel River. Parry decided to leave immediately to see them both for himself (See Map 5).

The journey may have reminded Parry slightly of his great discovery on the western side of Lancaster Sound. He now entered new land along the Hunter River which had been ignored by Dawson in his original survey, despite good soil quality, easy contours, shade and an abundance of water. The party of fourteen rode through the Murrurundi Pass in the Dividing Range and onto the plains on 13 March 1832, and on reaching Warrah they turned north via Quirindi to the Peel River, the closest of Dangar's two eligible sites. Parry wrote in his journal on 16 March, fully endorsing Dangar's opinion of this tract,

> It is *all* most splendid Country – not a single acre on which sheep may not feed. Apple trees in vallies [*sic*], Box on hills. Rich black deep soil in all the flats which are *forest* flats not like our alluvial flats. The Valleys are 10 or 12 miles long, and terminate in the plains. Where they terminate the plains are exquisitely beautiful.[6]

The following Friday, 23 March, the group set out for the second location now estimated to be around 250,000 acres and which, like the first, was outside the official limits of settlement. To reach it they had first to cross the Peel River swollen by heavy rain, and Parry improvised a lifting system based on naval 'sheer leg' principles with which they moved their baggage across in under five hours, leaving the oxen, horses and men to swim over. Non swimmers were transported on a raft of drift wood. Once across they were greeted by five aboriginal men who asked anxiously, after some small talk, whether they could have their hair and beards cut! Flour and tobacco were offered but rejected and Parry noted that they had no interest in alcoholic spirits either.

The land they had reached was unquestionably suitable for the Company's purposes. 'The Downs very much resemble the Liverpool Plains but have rather more grass mixed with herbs,' Parry wrote. They then turned west after fording the Peel River, to explore downstream to examine the country as far as the Murrurundi. He entered into his journal,

> The forest land recedes alternately on each side of the River affording excellent positions for alternate sheep stations. Our impression is that these pretty hills which have only distant tufts of grass, would be even more *healthy* for sheep than those about Warrah etc. but would not support so many.

The party turned south-east, crossing the river again near Wolloomal, thirty miles upriver from Tamworth. This land was to become known as County Parry on later maps, and had fine sheep pasture with beautiful and flourishing new grass after the recent rains; however, in extent it was only 150,000 acres and it had one serious disadvantage. The lack of any direct route for droving sheep and cattle down through the Great Dividing Range to Port Stevens. This difficulty meant them returning by way of the Crawney Pass, a longer route but by travelling hard they arrived back via the Gloucester River at Port Stevens on 16 April after an absence of six weeks.

Enthusiasm for the land they had surveyed prompted Parry to write to Bourke, the new Governor, asking that *both* tracts shown on Dangar's map be granted to the Company in exchange for a similar area from the current AAC holding. He made preparations to travel to Sydney immediately in order to petition the Governor personally, but on hearing that Bourke's wife had just died he waited for his summons. This came on 1 June, after a two long months of waiting which he found agonising. By 4 June he was staying as a guest in Government House where he composed a lengthy report to the directors in London stating that 'no real objection could be made, if the new lands were brought within the limits of the Colony which at present they are not'. He anticipated objections from the new Surveyor General Major Mitchell, who had been mapping the identical area for road construction just before his own expedition.

Bourke had been briefed by Mitchell about the current settlers' rights in both the locations and had launched his own counterproposal which would give the Company 600,000 acres on both sides of the Peel River with a boundary along the top of the hills. This was far poorer and more barren pasture land with little or no water. Long discussions took place during which Parry admitted he found it hard to keep his temper with Mitchell, a strong-minded man with a penchant for a good quarrel. Bourke rightly referred the matter back to the Government in London.

Although Parry was always prepared to make concessions, these proved unnecessary. After a year waiting for a decision it was agreed in accordance with his wishes, and in 1833 Lord Goderich, the Secretary of State for the Colonies, ordered the Governor to give the Company freehold to both parcels of land. 562,900 acres in total, displacing the 12,000 head of cattle and 3,000 sheep owned by landowners who, in many cases, lived elsewhere.

It was the culmination of his efforts to secure the prosperity of sheep farming in New South Wales. With less than two years of his four year contract remaining he could start to improve the quality and number of the sheep placed on the new grant land, and move the headquarters of the Company from Carrington to Stroud, which would be much closer to the epicentre of the Company's improved holdings. With James Ebsworth, his deputy and chief accountant, they could also overhaul the entire accounting and reporting system and he could spend more time visiting and superintending the various stations' shearing barns, and the development of Stroud with a new flour mill, sawmill, schoolhouse, and at his own expense a new chapel which was completed before they left Australia. This chapel is now elevated from chapel to a full church and visible from the Pacific Highway that passes through the township of Stroud.

Within months Thomas Ebsworth, James' brother, Charles Hall and Andrew Turnbull were busy at Stroud getting ready for the arrival of several thousand sheep and finishing the storerooms ready for the wool from the next shearing. Dilapidated housing was rebuilt as a home initially for Parry and his family, and subsequently for use by Henry Hall who would become the Company's agent and local superintendent. While they were away inspecting these projects Parry and James Ebsworth prepared a full set of consolidated accounts for the directors, forecasting that although between £50,000 to £100,000 (between £4 and £8 million) of capital at today's values had been squandered, they could see the *possibility* of a return of dividends arising within the next two years.

Parry went to Stroud again on 22 September to look at the school site and catechised the school children himself. He left remarking in his journal, 'my anxious wish is to get the whole establishment up here as soon as possible.'[7] Carrington would be nothing more than a 'wharf and a warehouse; the deserted village which I wish to see it become unless we can let the houses to emigrants or others'. His dislike of the township contrasted with his home of Tahlee nearby, where he and Isabella had dedicated themselves to the improvement of the convicts' and the natives' lives. The work they accomplished there consisted of keeping relationships calm by settling disputes between the English and Irish convicts, reducing high levels of alcohol consumption and the frequent drunkenness among the people and the Company's servants. They had to face many other minor disputes but none diminished their quiet enjoyment of the small settlement.

Their combined efforts to give local children a proper schooling and the employment of a schoolmaster resulted in steady improvement to both numeracy and literacy. Church attendance had become more widespread both through the baptism of infants and by encouraging confirmation leading to more than twenty communicants at the Carrington Church.

Some people came purely to follow his own family's good example and for his direct and robust style of preaching. The Tahlee population had become souls entrusted to his care in a way similar to his outlook towards his sailors. As well as looking to their spiritual needs he managed to found a local savings bank which eventually became the Bank of New South Wales. It was for all the paid employees and their families to encourage them to accept their responsibility for financial security in the future.

Isabella's social work and his relentless travel and site inspections over a huge geographical area, as well as the new business projects he had started, were now telling on them both. She weighed only eight stone and wrote to her mother that her hair had turned grey, while he admitted to his mother-in-law that, 'I am grown old, grey and suspicious of everybody.' He looked older than his 43 years with a deep furrow between his brows.[8]

Part of the Commissioner's work was regulation of the judicial system and listening to complaints made against the police and individual members of management. The cases were mostly instigated by indentured clerks and other company servants who came to believe that their actions were inviolate, beyond criticism or reproach, and subject to censure only by the board of directors in London. It is very clear from his daily journal that the convict population proved more useful, biddable and helpful because – if they were not – their hard-earned privileges might be removed.

During the last two years of his contract the number of convicts transported from Britain rose by 60 per cent, mainly due to the public protests, demonstrations and social unrest in southern England arising out of agricultural mechanisation and desperately low wages. This influx proved timely as the convicts included experienced shepherds, skilled tradesmen such as carpenters, wheelwrights and masons, and several trades needed for his ambitious plans at Stroud.

There was also a shortage of skilled men at Newcastle and Port Stephens. The professional occupations he failed to find were schoolteachers for the new school at Stroud.

At the end of 1833 Parry sent a long letter to his brother Charles commiserating with him over his acute depression. It contained a summary on the religious beliefs he held which had become embedded during recent years. He was hypercritical of an English educational system that removed religion from children of the poorer classes and the steady reduction of religious education in Britain, 'God is out of sight in the schools and in that reprobate and unchristian senate of ours.'

He pronounces in a simple statement, his belief in the efficacy of Christianity: 'It is a daily source of admiration to me how *exactly* suited to the wants of man both spiritual and temporal, the Gospel of Christ is.'

He finished with confirmation that he will return to England in the autumn of 1834 and would like, if possible, to buy a Welsh cottage. Perhaps this was recognition of his Welsh background and a desire for life with simple attractions in rustic surroundings such as those they enjoyed at Tahlee.[9]

For Christmas 1833 Parry arranged special celebrations knowing it would be their last in the colony, and took the unusual step of inviting some of the local aboriginal people to a bullock roasting, and laid on another, separately, for the convicts. The veranda at Tahlee was converted into a dining room for the customary dinner on New Year's Day for all the officers of the Company with their wives, then followed by games and races for the children.

The health of little Lucy began to cause alarm soon after these celebrations. On 11 February Lucy and Crowther, her nurse, went with Parry on the new steamer from Newcastle and reached Sydney in eleven hours, half the time taken by a sailing ship. His journal does not relate the treatment Lucy received, although a Mr Annesley, a specialist in tropical diseases and a member of the Medical Board of India examined her. She suffered a fit later in the year with a probability she was mildly epileptic, but not suffering from the amoebic dysentery common in New South Wales at the time. Two weeks later Lucy, her nurse and her father were back at Tahlee where he was again deluged with correspondence, petitions, and legal writs from Mr Barton against both the Company and Parry himself.

His final year of 1834 proved even more strenuous than earlier ones. There were many new ventures to supervise and different trial crops, including tobacco, corn and vanilla, to cultivate and record. Future plans included a salt pan near to Newcastle, trials of newly imported sheep breeds, cross-breeding between the various cattle herds, and one of Parry's own interests: the trial cultivation of different varieties of grape vines sent from Europe for wine production. These early attempts produced, after years of experimentation, the now famous wines from the Hunter Valley. On top of this work he continued sending full reports back to the directors every three months which, despite Ebsworth's expertise and diligence, was a huge administrative task keeping him up late into the night when he was at home.

In October 1833 Isabella had given birth to another child, a boy, who was named Charles, after Parry's brother. After the baptism, Parry sent the details to Charles and also expressed his sadness and grief at the news of the death of his great friend and mentor Sir Murray Maxwell two years earlier. A loss which he admits was made worse by regret for not having written to him more often.

At the end of 1833 Parry was preparing to hand over his duties to his successor, Colonel Henry Dumaresq, who had been living in the colony for some years. Dumaresq had been private secretary to Governor Darling and was widely respected. He was intelligent, had integrity and was the owner of a large estate at St Heliers in the Hunter Valley, along with two of his brothers.

Parry wrote in his journal that on 26 January that they all went to church in St Phillips Church in Sydney, having finalised the arrangements for the date of a ship to take them back home. This was the 46th anniversary of the founding of the colony.

During his final three months Parry continued to travel between each location where the Company had operations, checking results, adjudicating over disputes and grievances, and conducting many church services. At these he continually emphasised in his sermons the chief values that he stood by and especially those from Acts of the Apostles, Chapter 20 verses 32–35. His desire to leave the colony as soon as possible was shown by an emphatically worded journal entry of 28 February 1834. In this he retells a confession made by a man called White who had fabricated a report about his supervisor's drunkenness: 'Oh this horrible, this dispiriting place. God be thanked that I and mine are about to leave it!'[10]

After a dinner given for all the local officers of the Company and held in a large tent, the Parrys started packing up Tahlee. On 18 March Colonel Dumaresq arrived with his family and after a night of 'consultation' he endorsed the Letter of Attorney which was signed and dated, and his duty as Commissioner ended. The family, with Crowther, a new dog and several other pets, set off for Sydney at 6.00am on 19 March in the *Sophia Jane* steamer; they arrived at 10.00pm and went directly to Pulteney's Hotel. Here they were forced to stay until the 27th for cases brought by Barton to be heard in the Supreme Court. None of these suits amounting to a total claim of £2,500, (£170,000 today) succeeded, largely because Barton, obstreperous and self-opinionated as ever, decided to represent himself after his legal counsel withdrew. Parry had sensibly obtained subpoenas in advance from all the key witnesses.

This distressing last episode to his four years in Australia was compensated for by a lavish dinner followed by many complementary speeches given by officers of the Company based in Sydney, as well as Governor Bourke's own praise and gratitude. The following day Bourke and his wife took Parry and Isabella on a special tour in his carriage to Liverpool, a nearby town, showing them the new hospital, the gaol, the police barracks and the post office. In addition to these, the first stone bridge built in the colony! It was, Parry wrote with surprise and some pleasure, 'a neat country town'. His memory of this well-found settlement and the improvements he saw with new public buildings and churches may have convinced him at the very last moment, of the future success and prosperity for the new colony. Rashly, he decided to invest money in the Bank of New South Wales before they left and to join its board, rather than take all his savings back to England where he felt that financial disaster was imminent. As it turned out, the Bank of New South Wales – in which he placed much of his hard-earned bonus – collapsed, leaving him with little and facing legal charges from several shareholders.

Just before leaving New South Wales Parry wrote two letters which also reveal more of his feelings about his years in the colony; one was addressed to the Cresswell family in Norfolk and the other to John Franklin.

To the Cresswells he declared:

I have the satisfaction of the feeling that my exertions have been the means of cleansing an Augean stable which, even to myself, seemed hopeless, for the first two years of my residence in this colony. But it has cost me more than the four years I have spent here, having told on me like ten.

To Franklin he admitted,

You will I am sure be desirous of knowing whether my coming to this country has answered my expectations. In point of emolument, I answer that it has. But the country is so dreadful from a moral point of view, and the duties that I have had to perform have been so often painful, that I certainly should not have undertaken the office, had I known what it was. Still we have, I trust, been the means of doing much good not only to the worldly concerns in which we have been engaged, but, we hope, to the religious and moral wellbeing of our little community.[11]

He continued by giving painful details of his own and his children's health.

One of the most useful legacies he left behind was grape cultivation, both at Tahlee and in the Hunter Valley. These included the Semillion and Shiraz grape varieties. He visited one landowner, Sir John Jameson at the Regent Ville, but was unimpressed with the quality of wine he made and wrote candidly his view that, 'wine making must be a matter of speculation and experiment at first'.

His own special interest was one which, fortuitously for the colony, the AAC wisely decided to gradually develop over many years.

On 20 May the family and servants, accompanied by a large collection of animals including a cow, a kangaroo, an opossum, five sugar squirrels and various birds, boarded the *Persian* for the long voyage home.

Chapter 15

Unfinished Business

The Parry family, their servants and the large menagerie they had brought home, landed in London on 26 November 1834. Their intention was to spend some weeks at Alderley to recover from an unusually long voyage. The *Persian* was 'the most uneasy sieve I ever sailed in' according to Parry, and it had been a great trial for everyone, especially Isabella who was never a good sailor. After a few days at Audley Square, where Lady Stanley had sent a nurse to look after the four children, they set off for the Stanley estate via Manchester.

Areas of the country through which they passed were in near rebellion. The mechanisation of agriculture had created widespread unemployment and poverty; respect for property, an unbroken rural tradition for centuries, seemed to be melting away. 'Incendiary fires lit up the darkness of the hungry night throughout thirteen of the southern counties.' The demand for Parliamentary reform as well as repeal of the Corn Laws had formed deep divisions between rural and urban communities, landowners, the two political parties, as well as the rich and the very poor. Laws enacted centuries ago were no longer fit for purpose in a rapidly industrialising nineteenth-century Britain, and of these the Poor Laws were a notorious example.

Poor Relief had been administered by individual parishes in England and Wales since 1601. It had become seriously abused by some employers and was very costly. Landowners used it freely to substitute for wages that they could well afford to pay and this malpractice drove down the wages of both skilled and unskilled men. It also undermined the size of the agricultural labour market by reducing the number of men applying for work, since, as the parish support was widely available, there was less reason to work long hours for a relative pittance. Under the Law of Settlement an unemployed man was tied to his place of birth, and this subsequently created pockets of unemployed labour. The subsidised wages being paid out had increased by so much that it appeared as if the entire British economy could be threatened.

Parry arrived back in England three months after a new law had been passed which was intended to remove these difficulties. The new measures were created to force the able-bodied worker back into the labour market, and the designation of 'pauper' made more demanding by the workhouse test. Conditions in the workhouse were to be even more harsh than conditions outside, both as a deterrent and to limit numbers being taken into the workhouse to only the neediest. The parish relief would be set at levels lower than the lowest paid man in local employment. It would only be paid if a man and his dependants entered a workhouse where they would receive relief mainly in kind such as food and clothing, instead of money.

For the purpose of applying the new laws consistently all the many parishes would be grouped into local Unions with elected Boards of Guardians who would build workhouses and administer relief through paid officials. Assistant Commissioners would travel across an appointed region, mostly based on county boundaries, enforcing the new arrangements and looking for any false accounting.

The abolition of parish control broke a long tradition of autonomy exercised by local government and was greeted with deep suspicion and dislike. The new Act did not recognise that more than half of those in receipt of the old form of parish relief were the elderly, women and children, and the infirm or insane. Able-bodied men were relatively few in number.[1] Furthermore, it was recognised by the better off in society that the changes were intended to lower the birth rate among the poor.

As the Parry coach travelled northwards the signs of unrest were apparent although the countryside around Alderley was generally quiet. The Government had hesitated to enforce the new act in the northern counties from fear of unlimited rioting and of infuriating the large Whig land-owning interests. It was a sombre homecoming. The weather was cold and the children born in Australia, who had yet to adapt to the new climate, felt it badly.

Early in 1835 Parry travelled back to London to visit the Admiralty and to attend a testimonial dinner given in his honour by the directors of the Australian Agricultural Company. The dinner, to which he invited Barrow and several naval friends, was a success but there was no suggestion of a commission from either Barrow or the Admiralty. He had been asked to stand as one of the Tory MPs for Bristol, but realised he would be in contention with Henry Hobhouse, a cousin, and also that it would be ruinously expensive.[2] In the end, Hobhouse stood for Finsbury.

In a letter to Charles Parry apologised for being slow to visit either Bishopstoke or Bath. 'I feel unequal to any further excitement. I am much disappointed at this as I have a thousand things to say to you as well as hear from you. The state of my health requires more care than I have been accustomed to give it during 32 years of "vagabondizing".[3]

During his testimonial dinner he had been told that appointments were soon to be advertised for Assistant Commissioners of the Poor Law and one of them would be for the county of Norfolk. Both he and Isabella had many friends there and because of another serious outbreak of typhus in London, he decided to apply and asked Charles to solicit from his Whig friends a testimonial for him; 'a man of business, zeal, and integrity'. His financial position was serious, having left a considerable sum invested in New South Wales, but if appointed as a Poor Law Commissioner he would receive a salary of £600 per annum, a little more than £42,000 today, but his naval half pay of £190 pa would be stopped. If he accepted it meant a £400 improvement overall, which was 'worth my normal profession' and yet unfair, since the work involved, he was told was 'most laborious'. The most important benefit he foresaw was living in Norfolk, which would be cheaper and safer than London.[4] It was a decision that ultimately he regretted, for although his family enjoyed a healthy life in the countryside quite as peaceful as at Tahlee, he was rarely able to share in its simple pleasures. While accepting personally the need for Poor Law reform he became deeply concerned over the consequences wrought by the new workhouse rules.

He urged the commissioners to discourage the migration of families to the northern industrial cities where labour, especially child labour, was competed for by the Lancashire cotton mill owners. He urged that children should never be separated from their parents and that the manufacturers must sign employment contracts for a minimum of three years. Ten thousand families though did move north, many from Norfolk and Suffolk.[5]

His duties included tours of the whole of Norfolk and adjacent parts of Suffolk, encouraging the adoption of the new Act and the setting up of new Unions and workhouses. The new Act did not recognise the simple fact that in some places parishes had already formed voluntary unions under a previous parliamentary statute which had now to be set aside by voluntary agreement before new arrangements could be imposed. They had to be 'induced to commit suicide', as Parry wrote to Frankland Lewis, to whom he was directly responsible. His duties were as physically arduous as those in New South Wales and he dutifully visited all the towns and villages to explain each of the 110 clauses of the new Act, while trying to persuade the reluctant guardians to disband any old parish arrangements.[6] He met with stiff opposition from rich landlords, some of whom refused to believe the new Act applied to them, and – on more than one occasion – the Church of England. A pamphlet written by Reverend Ambrose Goode of Terrington expressed the opinion: 'To lovers of Liberty, I say, let us resist by all constitutional means this new Poor Law amendment Act as a violation of the fundamental law of the land in depriving the poor of liberty of person and as it makes poverty a crime.' Even *The Times* chose to describe it as the 'Starvation Act'.

These words must have been exceptionally painful for Parry, whose feelings lay consistently with the poor and disadvantaged. He called on the Reverend

Goode who, faced with Parry's determination and personal charm, backed down, withdrawing all his pamphlets.

The Act demanded the submission of all parish records of expenditure on the poor for the previous three years and Parry needed to act decisively and quickly before these were deliberately destroyed. Another part of the work he disliked was needing to negotiate with and calm rioters who burned down a workhouse in Great Bircham. This incident was so serious that he needed to ask the magistrates to bring in the Metropolitan police as well as the local coastguard. At the time Norfolk had no police service of its own, but due to his usual persistence this was eventually established in 1839.

In the City of Norwich he had an especially difficult time because of the scale of unemployment and political manoeuvring by the guardians of the large and influential existing Union. A majority of these men resisted the changes, being brewers or shopkeepers 'into whose pockets,' Parry commented, 'by a very common and natural process of circulation, the money, or a large proportion of it, finds its way.' A rough estimate of the Norwich relief was £30,000 (£2.1 million today) each year.

The first house they rented was at Mattishall near East Dereham and once had been the home of poet William Cowper (1731–1800). Parry had for years been an admirer and read his poems and used his hymns for Sunday services during his Arctic winter imprisonments. Cowper was an anti-slavery campaigner, a devout Christian, and wrote emotionally about beauties in nature and the countryside. Unfortunately his house reflected more of his melancholic side than of the latter, and Isabella found it dark and depressing. She became sickly and depressed and Emmeline, her sister, was sent by Lady Maria to keep her company and report back to their mother on her state of health. Lady Maria usually insisted that Isabella's poor health and her grandchildren's illnesses were the direct result of too close an association with 'pauper children.' Parry could not agree, but knew that because of his frequent absence he needed to find a house that she liked better and a setting that agreed with her.

Early in 1836 they moved to Congham Lodge near Castle Rising. This was a far nicer house with a productive garden and close to many good friends including the Cresswells, the Frys, Lady Middleton, the Cunninghams and the Hoares. A number of these friends being Quakers.

Isabella's health improved and she became pregnant once again. The children, however, contracted scarlet fever and Emmeline stayed on for weeks to nurse them. Most recovered but the youngest child, Isabella born in Australia, died on 6 March and Isabella now in mid-term, was heartbroken. This daughter, she felt, was truly part of herself and reminded her much of her own passionate childhood. The following May another girl was born and immediately christened

Emmeline. Parry wrote to his mother-in-law: 'The two elder children are delighted with it, Charlie dubious.'[7]

Parry's own health was by now deteriorating once again and he wrote to the Commissioners saying that he wished to resign. While awaiting their reply he wrote to Charles that he felt:

> It is comfortable in one sense but uncomfortable in another, to know that my success in this county has been considerable. It would certainly have been more agreeable to have reaped the full benefit of my exertions, but this cannot be helped, as I am so unstrung that I must have rest.

He had created ten new unions consisting of 280 parishes and, thanks to his hard work and that of his fellow commissioners, better local supervision and governance of an important foundation for the newly developing industrial system. The completion of the Reform Bill by the Whig administration allowed England to navigate through a period of dangerous social and political turbulence without serious bloodshed. The beginning of what some historians describe as, 'the Victorian age of balance'.[8]

During the hectic weeks finding his way around rural Norfolk, an unwelcome source of distraction and aggravation arose with the return of Captain John Ross. He had raised sponsorship from the gin distiller William Booth and equipped a second-hand packet ship called *Victory*, with new paddle wheels that could be lifted clear of the water to avoid ice floes, and installed a new steam engine from Messrs Braithwaite and Erickson involving bellows instead of a funnel. This engine had been inspected by Parry who described it as 'too new and untried and not [for] the sort of service on which novelties of that sort should be tried.' He had agreed though that the 'application of steam as the moving force to this project offered 'a very great hope of its accomplishment'.[9] Ross was single-mindedly determined to rebuild his reputation as a successful Arctic explorer and prove that he could find one route through Prince Regent Inlet into the Northwest Passage. Where Parry had failed, he would succeed.

His large crew included James Clark Ross his nephew as second in command. He having been on half pay after the North Pole expedition with Parry. Several officers who had also served under Parry offered their services without pay, but Ross petulantly turned them all down. Franklin and his new wife Jane Griffin went to see the *Victory* and public enthusiasm gained momentum again over the search for the passage. John Barrow would take no part in it and the Admiralty Board considered the ship too small and unfit for the work. Possibly to annoy Parry, John Wilson Croker his old critic from the Hydrographers office supplied Ross with a special Admiralty letter of introduction were he to finally reach

Russia. The expedition left the Thames in May 1829 and reached Stranraer in Scotland on 10 June. The new engine broke down frequently and they lost a foremast in a strong gale. The steam engine was found to be almost useless and they limped into the northern part of the Regent Inlet on 10 August with a single serviceable paddle wheel, and then disappeared for almost five years.

The published version of this voyage is a memorable account of their survival through four long winters, a lively relationship with local Inuit who were welcoming and helpful, and the exceptional endurance and dedication shown by the officers and crew. Ross behaved as before, drinking heavily in his cabin and emerging briefly only to issue orders. He once again took on the work of the expedition artist while James Clark Ross was in charge of scientific work as well as the organisation of sledging parties into the interior and round the coast of the Boothia Peninsula where he marked the position of the Magnetic North Pole, raised the Union Flag, and named the place Victory Point.[10]

After enduring two harsh winters with very low temperatures, the dismal and silent surroundings and a lack of fresh food and greenstuffs, not appearing to know about the uses of sorrel, all began to undermine his own and his crew's health. Ross decided to abandon *Victory*, sledge across Boothia to Fury Beach, collect more tinned food, repair and launch the three boats left behind by Parry, and return up the Regent Inlet towards Lancaster Sound. During this endeavour they needed to survive on two-thirds normal daily rations.

The stores still on Fury Beach proved to be their lifeline. Using wood from the wreck they built a large and secure wooden hut named, with obvious irony, Somerset House, after the Admiralty headquarters. They heated this with coal from the *Fury* and struggled through another dreary winter. The following spring they attempted to reach Lancaster Sound in their small boats but failed and returned to spend another winter at Somerset House. Nearing the end of all their rations they then again rowed north and reached Navy Board Inlet where they were soon picked up by the *Isabella*, the whaling ship Ross had commanded on his first expedition.

On his return home Ross immediately started his account of their voyage. He made exaggerated claims about his discoveries, levelled accusations against the manufacturers of the steam engine, appropriated for himself most of the credit for finding the Magnetic Pole from his nephew, and made little effort to explain that the Magnetic Pole was not the precise geographical location.[11] His remarkable talent for publicity and self-promotion were clearly shown in his deliberate adoption of eye-catching headlines for his newspaper articles including 'back from the grave' which held many people's attention. Now, he had no doubt, his position was strong enough to restart the twelve-year-long argument with Parry again, accusing him of 'gross misconduct'.

He needed Parliamentary approval for an award of £5,000 (£35000) that he had claimed from the State but Parliament was suspicious and critical questions began to be asked. Ross soon saw that his first controversial expedition was being compared by MPs with this latest one. One remarked, 'the expedition of Captain Ross was taken more with the view of recovering his reputation than with the view of benefiting the public by discovering the North Pole.' A Select Committee was formed to investigate and if, finally, he could clear himself from the criticism of his actions on the first voyage, it was far more likely to approve his grant. He therefore set out to improve his reputation as the latest and bravest Arctic hero, discoverer of 'The Pole', and to create public adulation and respect by every possible means. Their support with the newspapers he knew could force Parliament's hand.

Parry decided not to argue with him again, but expressed to Franklin his frustration and disappointment that the man could behave in this way. He was now accused by Ross of being in 'neglect of [his] duty' by failing to tell Ross the Croker Mountains didn't exist, and described as a 'rascal' for not telling him so.[12]

Franklin urged Parry to answer these 'shameless and studied misrepresentations', but he finally decided only to send his letter book, daily reports sent to Ross, and his navigational and positional records details to Barrow who then wrote a further damning article about Ross in the Quarterly Review.

Parry explained to Charles, 'pamphlet will begat pamphlet and an endless controversy will ensue, of which one half the world would not understand the real merits and two-thirds the other half would not care about it'.[13]

Ross was eventually awarded a knighthood, his desperately needed Parliamentary award of £5,000 which helped clear many of his debts, and was also paid large amounts for his voyage account. He was also given a diplomatic post in Stockholm quite possibly to remove him from the sight of the Admiralty! His voyage achieved almost nothing new and had mistakenly recorded King William Land as part of the North American mainland, whereas it was a separate landmass around the most southerly point of which was a narrow strip of navigable water leading out west towards the James Ross Strait and Point Turnagain, where Franklin had halted on his first overland expedition. This was a piece of the jigsaw which, had it been thoroughly surveyed, might later have given Franklin a different choice of route to follow and a slim chance of survival.

His voyage did however, produce two important and unexpected outcomes. By again seeking the Northwest Passage he had created a frenzy over his 'heroic' return from the grave and reignited interest in this near discarded project. This, ironically, gave the Admiralty help within Parliament for fresh funding for one more large Admiralty-organised expedition. Another was Ross' abortive use of steam which although a failure had indicated, though not proved, that this new technology in a stronger and larger ship using screw propellers instead of paddle wheels, might succeed.

Parry, awaiting his replacement as Poor Law Commissioner, spent more time at home although his contentment was blighted by the death from scarlet fever of his daughter Isabella born in Sydney. He had also decided to home school his son Teddy (Edward), and epileptic daughter Lucy chiefly to keep them away from possible infections. Just before Christmas 1836, after saying farewell to their many friends and neighbours, they packed up once more and went north to Alderley.

Chapter 16

Steam Machinery and Packet Boats

Parliament had decided to transfer to the Admiralty twenty-four steam packet boats run by the government postal services which meant that the number of ships driven by steam then used by the Navy doubled overnight. The post office used steam paddle ships to compete with the growing number of commercial steam powered passenger ships on routes between England, the Continent and Ireland. The postal service realised that only by attracting passengers could they maintain the size of their own fleet and provide the frequency and reliability of mail deliveries needed. Competition for passengers however, was fierce and they failed to find as many as were needed. Two Parliamentary Commissions of Enquiry were held and it was decided that the services might be improved with reduced financial losses if the Admiralty took over.

Parry came back to London on 3 January 1837, and the transfer was completed on 16 January. Once again he was working for John Barrow and the man appointed after Croker, Charles Wood, the first Secretary to the Navy Board and the future Earl of Halifax.[1] His role involved good organisational skills an aptitude for analysing different sea routes and viability and negotiation with shipping companies. Many new developments including the use of steam power were advancing in countries lately hostile or competitive to Britain including France and the USA. Becalmed on the St Lawrence River, Parry had seen for himself a steam tug pull his own ship, HMS *Niger*, up to Quebec, and quite recently the remarkable improvement to passage times between Newcastle, Port Stephens and Sydney, when the new steamer service began.

With his normal determination to 'see for himself' he visited and examined each of the different routes and harbours used by the postal packet service. By the end of January 1837 he had been to Liverpool, Holyhead, Kingstown (Dun Laoghaire) and Waterford, followed by Milford Haven, and to Scotland for Ayr and Donaghadee, both of which he rapidly decided unsuitable, recommending Stranraer and Larne instead. His visits eventually took him 2,800 miles around Britain. He decided that the mail services to Ireland from Liverpool should be put out to tender and started discussions with the City of Dublin Steamship Company, the only large steamship company capable and interested in the contract. These were difficult negotiations and finally reached the conclusion that their existing postal service could not be bettered and that he could make them profitable. Thwarted by his change of mind, the Dublin Steamship Company introduce a pricing war with the government service, and the Liverpool to Dublin postal packets suffered a sharp drop in revenue of £1,000 per month (£68,000 today).

This bruising encounter opened his eyes to the harsh commercial realities of dealing with the major steamship companies, and still be able to manage a reliable, regular and economical service for the crucial markets and overseas possessions and colonies. He became convinced that if he awarded companies such as the Peninsula Steam Navigation Company with the Iberian mail contract, and Cunard the fortnightly transatlantic service with enough income the results would be beneficial to both Government despatches and British postal customers. The two large contracts were worth £34,000 a year (£2,800,000) today and £60,000 (£4 million) a year respectively, and a West Indies contract £24,000 (£1,620,000). The detailed contract work required much legal scrutiny and a clear understanding of the risks and benefits to both sides. He supervised the details personally so that when Cunard found their obligations for both regularity and frequency on the stormy Atlantic crossing during the winter months almost impossible to meet, he was able to explain to the Treasury all the facts and persuade them not to demand penalties, or demand the binding future payments agreed under contract. The terms given for this concession by Government were a binding commitment from the Company to invest in more powerful ships and to aid Government by 'carrying the heaviest guns and to give power to the Government to hire them in time of war'. In 1849 a Parliamentary Select Committee took the view that the contracts, worth £100,000 (£7 million at today values), were extravagant but, in the longer term, the subsidies which they were gave shipping line owners a solid income with which to invest in better, faster and more reliable mail and passenger ships, which gave a large benefit to British companies competing in the global market. It also provided Britain with the extra capability in wartime shown even as recently as the extra transportation needed during the Falklands Campaign.

During the summer of that year, and having been away from Isabella and his family for over six months, he managed to join them at Penrhos for a few weeks on the pretext that he was needed at nearby Holyhead. His brother-in-law, William Stanley, had inherited the Stanley estates on Anglesey and became the Whig Member of Parliament for that seat. Parry was present with Isabella to celebrate on election night.

A month later they bought a house in Devonshire Place just off the Marylebone Road and close to Regents Park. While this was being altered they rented another property close by in Wimpole Street, and in this house on 23 December another girl was born, whom they named Emmeline, the replacement for their favourite daughter, Isabella who had died in September.

The Parrys were quickly invited into the hectic London social life but found, as before, the cost of running their new house, even with a modest establishment very high and to economise decided to remain in London for the summer and show the children all the sights, a decision strongly disapproved of by the Stanleys, especially Lady Maria, who expected them to return to Alderley as usual. During these summer months the family visited the Zoo, Parry was elected a Fellow of the Zoological Society, and to Vauxhall Gardens to see a hot air balloon launch, and then to Greenwich to call upon the governor, Sir Thomas Hardy, Nelson's Flag Officer at Trafalgar.

The reorganisation of the packet services and the new contracts took him just over six months to arrange, and afterwards both Charles Wood and John Barrow were reluctant to let Parry go back into retirement. They convinced the Board of Admiralty to let them have 'a Naval Officer of skill and experience and with some scientific experience for the purpose,' (to be the first Comptroller of Steam Machinery).[2] Both men believed that his personal qualities, determination, objectivity and complete authority would result in only those new proposals significant and practical enough being submitted to other senior naval officers and Navy Board members; his recent experience with the steam packets giving him some knowledge and relevance of the new inventions both economically and practically. There were very few officers at the time who had experience of steam powered ships. A hardcore of Admiralty Board members, including Viscount Melville unlike his father, regarded steam power as suitable only for tugs, despatch carriers or ferrying supplies and crews between the large sailing men-of-war. It was not suitable for fighting men-of-war. especially 'line of battle' ships.

The exponents of steam power by this time were neither few nor short of any prestige and authority. In 1815 Lord Melville, then the First Sea Lord directed the Navy Board to build a steam sloop, to be called *Congo*, with a Boulton engine, to be used as a trial ship.[3] The *Regent* paddle steamer was run successfully between Margate and London and the *Comet*, with an 80 horsepower engine, was launched in 1822 followed by the *Lightning* in 1823.[4] While these two were built, HMS *Monkey*, a paddle ship powered by steam engines built by Boultons and Watt & Co., had been launched in 1821. She had been promoted some years earlier by Sir Joseph Banks as a shallow draft steam vessel for his naval expedition to the Zaire River.

More recently, the Duke of Clarence, Lord High Admiral of Great Britain from May 1827 to March 1828, took up the case for steam engines and forced a reluctant Admiralty to allocate a budget to their development. Predictably, it was Croker, then still First Secretary to the Board in agreement with Admiral Cockburn, who stood firmly in the way of engine research, new boiler models, propulsion methods and discussions with the leading engineering companies. Both men were convinced their views were right.[5]

Another individual standing in his path was Captain William Symonds who had been appointed to the office of Surveyor of the Navy in 1832 and dominated British naval shipbuilding design for many years. He introduced several new and innovative sailing vessels but was 'unprogressive regarding steam machinery and its adoption'.[6]

The position of Comptroller of Steam Machinery was created to allow development of steam engines, boilers and their transmission systems suitable for warships, but Parry was not allowed to become involved with the overall design of the ships themselves. Nor the training needed for the new engineering complement. This made the position awkward, since any new machinery was sure to influence both the ship's design and the number of engineers needed on board. Some of his opinions stirred up comments from reactionary elderly officers, many of whom held senior positions in the Admiralty and were strongly attached to sail as the best means of propulsion.

After Sir Thomas Hardy became First Sea Lord, replacing Cockburn, a far more practical outlook prevailed which helped improve the positivity of several of his fellow Board members. 'He was not only a reformer but a most prudent reformer and seemed to behold a prophetic vision of the mighty changes which science and steam would effect in the naval service.'[7] Unfortunately for both Parry and the Royal Navy his tenure was very brief.

One decisive factor for future naval policy was the success of Brunel's *Great Western*, which in 1837 crossed the Atlantic in fifteen days. The first vessel to do so under steam power alone. This crossing was transformational, even though it showed up the many problems created by using steam-powered engines in conventional wooden hulls.

Parry became involved with a rapidly growing number of ideas and projects promoted by the larger and more successful marine engineering firms including Maudslays, Penn, Seaward and Miller, Boulton and Watt, Napier, Fawcett and Preston, and Rennie. His job was to shield Barrow and the Sea Lords at the Admiralty from a 'host of speculative inventors whom it is not always easy to satisfy or get rid of, especially when championed by officers of high rank'.[8] He was aware that the majority of Members of the House of Commons considered the Admiralty had an undue 'thirst for novelty', and always favoured the established Thames yards at the expense of Clydebank or other northern shipbuilders. Successful vessels powered by steam had been built by Caird and

Co in Greenock, Napier and Sons in Govan, with specialist boiler-making equipment in Merseyside by Lairds, which was very advanced.[9] Parliament had its own reactionaries and sceptics who complained about several innovations. These they perceived as either too dangerous, too expensive, or simply unnecessary, mostly without any sound understanding of the engineering involved or the significant advantages to performance they might provide.

Sir James Graham (1792–1861) took over as First Lord of the Admiralty from Hardy in late 1835, still with the new First Secretary, Mr Charles Wood. Both individuals were known for their tact, patience, good temper and modernising ideas, but despite this their ideas and reforms were introduced long after they had both left office. The pace of change was painfully slow.

Parry was impressed with one invention especially: the direct engine with the shaft positioned over the cylinders as opposed to working with 'side' beams and levers. The new engines for the same gross weight produced 100 horsepower more than the 'side' engine. He argued consistently and persuasively that 'tenders for marine engines could not be disposed of in the same manner as an offer for beef or pork for HM Navy,' as well as seeing the benefit of giving specialist firms a realistic profit which could then be ploughed back into further development.

By 1843 the size of the boilers, weighing as much as 100 tons, were still only capable of pressures of 15lb per square inch. Parry knew that this was the limiting factor to creating enough power for major fighting ships and for a naval steam fleet to become reality. Working with the distinguished engineer Joshua Field they developed with a few engineering companies, new tubular boilers with greater efficiency, compound engines, condensers, and other substantial improvements. These were then put onto the 'paddle wheel frigates' constructed during the years between 1840 and 1850 and usually less than 1000tons DWT. Several of these had engines built on the Clyde which proved to be mechanically efficient and reliable, including the *Vesuvius* and the *Stromboli*.[10] In 1841 one ship, the *Driver*, circumnavigated the globe while another, the *Devastation*, was the first to be fitted with Maudslays' twin-cylinder direct-acting engines using double storied flue boilers. The same type specified by Isambard Brunel for his *Great Western*.

In 1844 the decision was taken to convert an existing sailing frigate into a steamer, and the *Penelope* was cut in half, lengthened by 65ft and fitted with a single 'Gorgon' engine of 650 horsepower. This ship, renamed the *Gorgon*, was a hybrid of sail and steam power and considered to be the future of British naval design since, along with the new machinery weighing 435 tons, it could also store 500 tons of coal and carry forty-two guns. The steam frigates continued being laid down until 1851, the last being HMS *Valorous* (1,258 tons), the largest ever built. Among all the various innovations Parry oversaw he did not fully recognise the potential of the screw-type propeller. A Swedish engineer, John Ericsson, and an Englishman, Francis Petit Smith, used an Admiralty barge in April 1837 to demonstrate the benefit of their propulsion method, although fifty patents had already been filed for similar designs. After a trial with several members of the Board of Admiralty aboard, Parry concluded that it was excessively noisy and vibrated too much to be of use. Ericsson moved to the USA in 1839 believing that his invention would be rapidly installed in a US navy steam frigate, but caution on the part of the US navy prevailed and eventually it was used in the *Valhalla*, a merchant ship. Meanwhile, Pettit Smith remained in England working on a screw-thread propeller funded by the bankers Wrights and Co, who supplied venture capital for his Ship Propeller Company. He also maintained contact with the Royal Navy and two-and-a-half years later, in 1839, his ship the *Archimedes*, powered by a more efficient but no less noisy or rumbustious propeller, repeated the same cruise down the Thames. This time full of engineers with stopwatches!

Parry wrote afterwards to Sir Charles Hall his immediate superior at the Admiralty that:

> none of the Board (of Admiralty) could attend but there were on board a dozen gentlemen who are screw proprietors (i.e. shareholders). The first thing that strikes you is the abominable annoyance created by the sound and the vibration. It is like a most noisy flour mill as regards the former and the shaking very great. The whole vessel is occupied by the cabin, the machinery and boiler and a small place forward for the crew.

What still disconcerted Parry most was the lack of speed. At full throttle the *Archimedes* only made 7¾ knots. The fault lay mainly in the capacity of her boiler and her hull design both of which limited the effectiveness of its screw propeller. Despite these problems he decided that she might be a better and faster ship if sails were combined with an engine, and a modified *Archimedes* was sent on voyages in 1839 and 1840 circumnavigating the British Isles. She took part in the Naval Review of 1853 and was deployed during the Crimean War. Petit Smith, whose perseverance and 'untiring energy' had created this engineering success for Britain was rewarded with £6,000 (£365,000 today) from the Admiralty who had breached his patent, a £200 civil list pension and a knighthood, but his Ship Propeller Company collapsed in 1840, taking the Wrights and Co bank with it. Parry, feeling a degree of personal responsibility for its failure raised a sum of £3,000 (£190,000) for the family from his private list of engineers, including Brunel, the big steamship owners, shipbuilders and naval designers.[11]

It was several years later before the Admiralty, hotly contested by Symonds, got enough Treasury money for another trial propeller ship and the *Rattler* was built. The famous Isambard Kingdom Brunel was special consultant.[12]

During his final years as Comptroller, Parry worked with the brilliant engineer, Thomas Lloyd, later to become the first Engineer in Chief of the Royal Navy. He set up the trials in April 1845 for the famous tug of war contest in the Thames between the *Rattler* with its screw propeller, and the paddle wheeler *Alecto*. A contest which the *Rattler* won convincingly by easily towing her astern at 2½ knots.

Amongst his final work on this transition from sail to steam he built a strong case with Sir William Parker (1781–1866) for a new naval engineering works to be built at Woolwich so that engines could be repaired at a naval dockyard rather than sent at a high cost back to each manufacturer. He also lobbied hard for improved training for and many more engineering apprentices and set up a committee, chaired by his long-time associate in the Comptrollers office Joshua Field, to decide what curriculum was required. By 1844 the Navy had started a dockyard school for 200 boys teaching the practices and skills needed as dockside workmen and fitters, progressing to be ship engineers if they showed promise.

He sensibly foresaw new problems created by the steadily increasing number of steam vessels, both naval and commercial sometimes travelling by night and so recommended navigation light rules. All ships would need to carry 'particular lights arranged in one uniform manner', so that 'at first sight every seaman shall be able to say in what direction the steam vessel is going,' as well as the adoption of a universal 'rule of the road leaving no doubt on which side vessels were to pass each other and which had to give way under particular circumstance, which might otherwise involve a doubt'. Trinity House approved all his proposals and they became Maritime Law in 1840, by which time they had become vitally important, most especially in confined waters and rivers.

By 1846 the Royal Navy started expanding their steam fleet, and in that same year 120 steamships were registered on the Navy List of which twelve were propeller driven. Commercial steamships, however, had multiplied ten times over since 1835, so that there was 95,000 DWT (deadweight tonnage) shipping worldwide. By 1871 this had risen to 1,290,000 DWT.

Chapter 17

Departures

The move back from Norfolk had been essential for Parry's work at the new postal packet services office in New Street and at Somerset House. Initially, his time had been spent travelling to each of the postal packet ports around the country to see the various engineering works on the Thames and to make himself known to the big steam shipping companies in London, Liverpool and Dublin. This meant absence from Isabella and the children for many weeks, so she had moved back to Penrhos on Anglesey. His own health had improved during the two years living in Norfolk, but Isabella's condition remained precarious and Penrhos suited her well; the noisy streets and heavily polluted air of London were not conducive to convalescence, even in the genteel neighbourhood of Devonshire Place. Since the death of her beloved daughter Isabella and the birth of Caroline in 1837, she had been weak and in poor spirits. Nevertheless, she joined a charity in Welbeck Chapel which visited, and provided relief for, the poor: 'approaching the work with few illusions and much love'.[1] This occupied her time but did little to improve her physical or mental condition, and with the prevalence of disease throughout the city, especially among the poor, she took considerable personal risks.[2]

By December 1838 Isabella was pregnant once more and he decided that the following spring the family would move away to a quiet secluded house in Tonbridge Wells called Mount Edgecombe. There Isabella might gain extra strength before her confinement, and he would join them all at the weekends. Doctors had expressed concern about her condition, especially the position of the baby in her womb and the swelling of her legs and ankles, but during her long rests she continued to write cheerful letters to her mother at Alderley, never mentioning the pain she was suffering.

On 11 May 1839 twin boys were born four months prematurely and lived for only a few hours. Isabella lost a large quantity of blood and was unable to speak. Parry sat by her bedside for two days and nights bathing her eyes and temples with eau de cologne and ether, but she did not rally and died on 14 May. She was not yet 38. He had written only the day before, 'I believe her life hangs on the most fragile thread: but I know that the dear Redeemer's everlasting arms must be around His own child and that it is well'. Nothing apparently shook his belief in the wisdom of God's judgement and the consolation he felt he could genuinely express at his beloved wife and several children being taken from him into Eternity.

Lady Stanley was told immediately of this terrible news and she arrived from Alderley the next day. She came with William, her clerical son, and within hours of Joseph Martineau, Parry's brother-in-law. Lady Maria never the most sentimental or sympathetic of women, wrote to her husband her thoughts:

> We arrived here at 4 o'clock and found a note from Mr Martineau to say it had been the greatest comfort to dear Edward (Parry) to hear we were coming. William went first, and returning with Mr M., we all went together to the Cottage. Such a quiet spot – and who could feel any painful excitement in seeing that dear creature's lovely face. It had not certainly the freshness of youth and health, but more sweet, more placid and more engaging it has never appeared – the features not at all drawn. I looked at her till I could have believed that I saw her breathe and that she was about to speak and the poor babies one under each arm — one a most beautiful child with well-formed features and the other not so well looking. They had not been baptised tho' they lived 10 hours, he had not leisure to think of it, but no difficulty is made or thought of as to their being buried with their sainted Mother. Dear Edward is astonishingly composed, with occasional bursts of grief. Much more comfortable to witness, than the cold rigidity of his features in general while repeating in his fine deep tone every trifling particular of her latter days, which gives him comfort to dwell upon. She seems to have been unable to converse, even to speak for several hours before her death, but kept her consciousness to the last with occasional delirium but of a very gentle kind. The medical men said they did not think it was painful to her. Who can judge?'

Her daughter Louisa took on the duties of mother to Parry's children and she wrote to her sister Emmeline, 'The four children Edward and Caroline, Lucy and Charles are all perfectly well and so much improved in strength health and beauty, nobody would believe that they are the languid creatures we had at Alderley. They are by far the most beautiful of our many nephews and nieces.' She observed that their father was normally 'composed', which helped keep their spirits up. Isabella was buried in the churchyard at Holy Trinity Church, Tunbridge Wells.

Parry felt his loss keenly and although unused to showing deep emotion other than to a few members of his immediate family, especially Charles and Caroline, he expressed his certainty that Isabella died because of his own sinfulness and made the same confession in a letter to Emmeline, who had nursed Isabella and the children in Norfolk and was the Stanley relation he felt closest to. He described his total desolation and loneliness as a physical ache and was convinced it was inflicted on him as, 'an utterly unprofitable man in the sight of God'. This profound observation he then moderated by words harking back to his successful and challenging years in the Arctic. He felt clearly, at this point, it was his main contribution to the world and his heart thereafter would lie with the ships and men of his polar voyages who represented his other family.

No doubt he also regretted some of the years spent in the Arctic wastes distant from his family and friends. 'I took dear Ted [his son] down to Chatham last Thursday to see James Clark Ross' ships before they go, and my old North polar companion now going towards the South.'

In this comment there is a note of regret that he is not going with them. His friend James Clark Ross' Antarctic voyage may have seemed an opportunity to escape the responsibilities of his family without Isabella beside him. But he was now too old and too busy on other Admiralty work to offer to go – besides, John Barrow was highly equivocal about this particular expedition. Colonel Edward Sabine had addressed the British Association for the Advancement of Science in August 1838 and had pointed out the deficiencies in their study of magnetism, particularly in the Southern Hemisphere. The Royal Society were much in favour of an expedition for this reason and wanted to re-examine the Antarctic continent scientifically after James Cook's perilous double circumnavigation seventy years previously which had left many questions unanswered.

After Louisa Stanley left Tonbridge, his own sister Gertrude came to Devonshire Place from Bath to look after Parry's four children. She was now 50 years old, unmarried and inexperienced with children. Although Parry was pleased to have help the arrangement was not a success. Teddy, now a boarder at Cheam School, remembered these years as profoundly unhappy.[3] Parry had decided, during these first desolate months of loss, to finish writing a book called the 'Parental Character of God'. These were his personal thoughts about the parental character of God as a Father and he asked his children to find verses in the Bible connected to this theme. Teddy believed that this was a way of reconciling himself to his loss and to learn how to become a better parent in the future.[4]

Gertrude decided that Miss Chenery, the children's governess, was not educated well enough for the role, even though she was ladylike and presentable. She suspected that perhaps the governess also represented some form of threat to her beloved brother's bachelorhood. Relations with Lady Maria became strained when she imposed her opinions, very robustly as usual, on the choice of governess; she defended Miss Chenery and this awkward situation was only resolved by the appointment of an extra French tutor; a cost that Parry could not easily afford.

Outspoken as usual, Lady Maria gave her firm opinion that he should remarry and suggested that he should see more of Mrs Samuel Hoare, who had been a close friend of Isabella's during their two years in Norfolk and was recently widowed. Catherine Hoare was part of the local Quaker circle that they had found friendly and convivial, and in sympathy with his feelings towards the less fortunate and under-educated in society.

Reverend Robert Hankinson, her father, was the Vicar of Walpole St Andrew and had been at Cambridge with the Reverend Charles Simeon, where he was deeply influenced by that man's evangelism. Catherine had three brothers, Robert, Thomas, and the youngest, Edward, who each took Holy Orders, and one sister, Mary Anne who was married to another clergyman: the Reverend Richardson Cox. The family was close and bonded by their dedication to a special kind of pastoral care for all their parishioners. The embodiment of a new and enlightened type of clergyman very much in keeping with Parry's style of evangelism.

At 20 Catherine had married Samuel Barclay Hoare, son of Samuel Hoare Snr and Louisa Gurney, but he died two years later, leaving her with two young daughters, Priscilla and Louisa. She then moved to Hampstead where she lived with her father-in-law, old Samuel Hoare, a banker.[5] In March 1841 Parry and Catherine were invited to stay with the Frys at Upton, near Tilney in Norfolk, and decided to marry and unite their two families. They were married three months later on 29 June 1841 at Walpole St Andrew.

Catherine Hoare was graceful, pious, dignified, musical and extraordinarily serene. She inherited her strength of character from her mother Elizabeth Gurney, a famous educationalist author and prison reformer. Parry's four surviving children combined in ages and temperament very easily with Catherine's two daughters, and she became a loving mother and stepmother as well as a devoted wife. They had three more children, even though he was in his early fifties and she eighteen years his junior. Their children included twin girls, born in December 1843, followed by a son in 1845. They were named Kate, Bessie and Willie. It was a happy marriage according to Teddy, his eldest son, and the children flourished. The constant health problems of delicate Caroline and the epileptic Lucy were treated calmly and efficiently and improved steadily while they mostly avoided illnesses such as measles, scarlet fever and whooping cough.

Soon after their marriage Parry was given an urgent new commission to inspect and report upon a scheme to improve the Caledonian Canal in northern Scotland. The waterway had been constructed by James Watt in the eighteenth century, with new locks and bridges engineered by Thomas Telford added early in the nineteenth century. A Parliamentary Select Committee of 1839 had proposed expenditure of £200,000 (£12 million) for further improvements such as deepening the shallowest sections, placing steam tugs on some reaches to pull sailing vessels through rapidly, and many more dredged anchorages with buoys. The Treasury objected to the high cost and asked the Admiralty for their opinion, especially about increases in traffic revenues they could forecast, which if realised, might cover the capital expenditure and recover the original investment of £1,000,000 (£60 million today).

In his report Parry recommended that the twenty-two miles of canal, twenty-nine locks and forty miles of lakes would in future be used mainly by steamships, not by sailing vessels since the navigation was narrow, winds light and fickle, and anchorages still few in number for much more traffic.. He recommended extra safety features at each entrance to the canal at Fort William and Inverness, and proposed that these all should be started without delay. For several years shipping did increase but, as steamships became larger and more powerful, the canal and its lock capacity became too small and they chose the long way round Cape Wrath instead. His report was published and debated by Parliament in 1842, just as Parry and Catherine moved from the house in Devonshire Place up to Hampstead Heath to be closer to old Samuel Hoare. The improved canal navigation was reopened in April 1847.

Parry was still nominally head of the new Steam Department and the Steam Packet Service but badly wanted to spend a quiet retirement with his young family, now consisting of six children. His health had deteriorated due to recurring kidney stones and gout, but his finances were in a poor state following the collapse of the Bank of New South Wales into which he had placed most of his Australian savings. He need to receive a regular salary – despite Catherine having an income provided by her previous father-in-law – was important. Despite their considerable wealth, the Stanleys did not, it appears, offer any financial support for their grandchildren's education.

The Stanleys, and Lady Maria especially, all practised the niceties of 'influence' and 'interest'. On several occasions Parry had to resist their attempts to place their appointees into positions over which he had authority, or to practise measures which would benefit themselves or their friends.[6] Relationships were not especially cordial with the senior members of the Stanley family, but the younger members were always his friends. Matters deteriorated badly when Parry heard about the shameful treatment his old friend Sir John Franklin had received from Lord Edward Stanley, the Secretary of the Colonial Office.[7]

Sir John and Lady Jane had found the same ruthless self-interest and corrupt behaviour in the Tasmanian administration, especially the officials appointed by the previous Lieutenant Governor – several of whom were related to him and were used to unscrupulous self-enrichment and a strong sense of entitlement, and were consequently resistant on principle to any change.

They opposed even modest attempts to introduce culture to the Colony, create new schools or to run the public finances in an open and honest way. The opposition press were hostile in the extreme, being especially spiteful towards Lady Franklin, accusing her of being, the 'petticoat' Governor in place of her husband.[8] Although most of the policies Franklin followed were decided in London by Lord Glenelg, Secretary of State for War and the Colonies (1778–1866), he was blamed when the Colony went through a deep economic crisis caused by a sudden collapse in the value of land and demand for Tasmanian farm products. This resulting from a decision reached in London in 1841 to stop subsidising any new settlers to Tasmania and to use it in future solely as a penal colony.

The most difficult individual was John Montagu, Franklin's Colonial Secretary, who had assiduously acquired land, money and political influence. He disliked Franklin's liberal and reforming approach and his 'interfering' wife even more.[9] Refusing to reinstate a corrupt official closely connected to Montagu, Franklin dismissed him as well and Montagu then wrote a highly critical letter to Lord Stanley impugning both the Lieutenant Governor's abilities and his sense of honour. Stanley, rather than listening and requesting both parties' versions and opinions, interviewed Montagu in London, accepted the version he provided, and wrote a report that reflected poorly on both the Lieutenant Governor and his wife. They were then recalled to England arriving in 1842, but Lord Stanley refused to meet him or even acknowledge that they had been treated very shamefully. Franklin possessed a strong sense of Christian duty and never shied away from confronting individuals of any rank who showed behaviour, or obvious self-interest that prejudiced fair and honest government. He made an enemy of Stanley by keeping to his position. Franklin took people at face value, trusted them as fellow Christians and on this occasion paid a high price for his morality and honesty.[10]

Parry sympathised greatly with Franklin, having suffered years of hostility and aggravation at the hands of several AAC officials in New South Wales, but he could not change the difficult situation by demanding his friend's reputation be restored.

A difficult enough task for anyone, but especially as he had often refused to comply with all the suggestions, hints or demands from various Stanleys for the preferment of people they knew. He was quite unable to lobby Lord Stanley for an apology and the redress that the Franklins deserved. Angry and dismayed by their treatment he did resolve to help if any future opportunity arose, which explains the complete support he gave to his friends selection as commander of a large new expedition to finally complete the last section of the Northwest Passage.

With his retirement from the Admiralty approaching, Sir John Barrow hoped for one final effort to secure a completed Northwest Passage for Great Britain. Again he lobbied for the support of the Council of the Royal Society, whose worldwide programme of research into magnetic variance of the compass led by Sabine was nearing completion. The Council, in their meeting held in January 1845, then proceeded to request Government for money again promising 'accession of geographical knowledge', and arguing that 'if such an expedition was deferred beyond the present season the important advantages now derivable from the cooperation of the observers with those who are at present carrying out a uniform system of magnetic observations in various parts of the world, would be lost'.[11] Barrow's scheme was to use the two vessels, *Erebus* and *Terror*, recently returned from Antarctica with James Clark Ross, to be commissioned with the best and most experienced Arctic officers and crews available, fitted with steam engines and sent to discover a route between Banks Land, sighted and named by Parry in 1819, and the north west coast of America; a distance of 900 miles.

The proposal was initially sent to Lord Haddington, First Sea Lord (1780–1858), with an attached note about a possible Russian expedition, intended to encourage a rapid decision. 'It would be most mortifying and not very creditable to let another Power (Russia) complete what we had done.' Haddington's response was to send the proposal to the most experienced Arctic explorers alive for their opinions: Parry, Sabine, Back, James Clark Ross and Franklin, but not Sir John Ross. Franklin answered enthusiastically the same day his copy arrived, replying that he considered there were other alternative routes should Barrow's suggestion be blocked with ice, and that this expedition would succeed. He wisely included a mention of the value of more precise magnetic readings and acknowledged the big new advantage provided by steam power with a single word: 'indispensable'.

Parry had written to Barrow and Melville after his third voyage to Prince Regent Inlet, saying that he was sure a route existed and would eventually be found. He could not now refuse to support the idea of a further attempt, but knew the extent and depth of the pressure ridges in the Beaufort Sea and the ice strength guarding the way down the Prince of Wales Channel south of Banks Land. He knew it would ned to be an exceptionally warm summer for this thick ice to melt sufficiently to let any ships pass through.

Several of the best candidates were unavailable. James Clark Ross was still recovering from three years in Antarctica and was soon to be married and declined for personal reasons. Parry and George Back were not well enough, and the next generation of officers such as Francis Crozier and James Fitzjames and were not felt by some at least to have enough experience to take overall command. At 59, Franklin was the oldest among them all, but his great enthusiasm for this long-awaited opportunity outweighed in his mind all the difficulties. He was overweight, had not been in the Arctic for seventeen years and his experience had been leading expeditions overland rather than negotiating the unpredictable and hostile sea ice conditions of the Polar archipelago. Parry, when asked for an opinion about Sir John's suitability perhaps recalling times when he had successfully manipulated his Admiralty connections at Franklin's expense, told Lord Haddington, 'if you don't let him go he will die of disappointment', and, more unequivocally, 'he knew no fitter man'.[12] But there were several sceptical onlookers, including Dr John Richardson who had travelled with Franklin overland on two previous occasions who knew his health was much weaker than it needed to be. Lady Jane also expressed doubts about his age and physical condition. Sir John Ross, who had much recent experience because of his most recent four-year voyage, was rejected completely, and Sir John Barrow who preferred James Fitzjames at 39 years of age, the same that Parry had been when he discovered his route through Lancaster Sound, was outvoted.

The decision to appoint Franklin was eventually decided chiefly on the advice of Parry, Haddington and Lady Jane Franklin – despite her considerable doubts.

As well as urgency on Barrow's part personally, and the need for final observations for Sabine and the Royal Society, there was also increasing concern that the Peel government might fall from office or simply change its mind. Preparations were therefore hurried. *Erebus* and *Terror* were strengthened to resist the heavy pack ice with fresh 10-inch belts of heavy timber along their sides and massive beams below decks on which to mount the two heavy railway engines and all their coal. The weight and displacement of the two ships was consequently considerably greater than before.

With his knowledge of steam equipment Parry must have known that the two adapted railway engines could only develop 25 to 20 horsepower each, giving a maximum speed of 3 mph even in calm ice-free waters. They would be unlikely to be able in the two heavily laden ships to push through even moderate pack ice. Even twice this power would be questionable and running both engines on full power would most likely cause their boilers to explode. The retractable propellers were an improvement over paddle wheels, but without sufficient driving power they could not make much difference.[13]

A new innovation was the desalination device that would improve the supply of fresh water but added yet more weight, as did their large supplies of coal, canned food, books, including Bibles for every man, and rum to last three years.

Only one officer, Francis Crozier, had had experience of steam power, having been on the Congo expedition on a paddle steamer.

Barrow had suggested to Lord Haddington that the 600 miles left unexplored could be undertaken over one summer because of steam power, a prediction made for the simple reason that it would keep Parliament quiet and ensure that the costs would appear affordable. Franklin and Parry regarded this as extremely unlikely and made sure that the ships were stocked for at least three years.

Franklin not unreasonably, wished to investigate for the passage starting from the west side and moving east, following the same Arctic coastline which he had fully explored and mapped. Using this route he had knowledge of the conditions such as prevailing winds, shoals and the landmarks and sheltering places in the immediate hinterland. His plan was flatly refused by the Admiralty with even Parry's backing, as well as the Colonial Secretary, who was none other than his old adversary Lord Edward Stanley.

One very outspoken critic was Dr Richard King, who recommended that two small boat expeditions should be sent down the Coppermine and Great Fish Rivers to act as search and rescue parties should Franklin's expedition meet with disaster. He published correspondence between himself and Lord Stanley in which he predicted that, 'Franklin was being sent to form the nucleus of an Iceberg.' Parry, with Barrow and others, dismissed his ideas out of hand since it reflected they felt on both the choice of Franklin as commander and the ultimate success of the venture. There were already many expensive preparations underway all paid for by the Treasury.

Interest and debate amongst the British press reached its climax on the day of departure of the two ships. Catherine described the visit paid by a family party of twelve to see the ships in dock a few days before they left.[14] Franklin gave them a guided tour, while Catherine gave Franklin copies of her brother Tom's sermons and some poems, as well as a drawing of Parry to hang in his cabin (see Fig. 19).

When the day arrived it was a bigger occasion than had been held for any of the earlier expeditions. Large crowds lined both banks of the Thames over long distances. On 19 May 1845, barely three hectic months after Franklin's appointment, the ships left the estuary of the Thames never to return.

Chapter 18

Tragic Hero of Polar Exploration

The disappearance of Franklin with all his men, and the many search expeditions sent by Britain and America to find them, has been well-documented by historians on both sides of the Atlantic.[1] There are also extensive and well-researched biographies of this courageous and remarkable man.[2]

His correspondence with Parry continued up to the date he left Western Greenland on his last voyage. In a letter dated July 1846 from the Whale Fish Islands, he announced that they had arrived in good time, the weather and the ice were favourable, and all the omens seemed 'propitious'.[3] As they were preparing to leave they met two whalers and Franklin spoke to a Captain Martin who said later that Franklin told him they had food for five years which could be made to last for seven. This was, if reported accurately, untrue, since they only had provisions for three years which might at the very best, be stretched to five.[4] Franklin's orders were:

> go to Cape Walker and then attempt to sail south and west. If that direction was blocked to attempt to sail north up the Wellington Channel to look for an *open sea route*. Failing that to follow up any clues or expectations which your observations may lead you to entertain.

The inclusion of a navigable sea route northwards up the Wellington Channel strongly suggests that Barrow still remained convinced that an open Polar Sea existed, despite Parry's North Pole expedition and all Scoresby's pronouncements.

Barrow had become obsessed with this idea and his influence in the Admiralty, together with that of the First Sea Lord Sir James Graham (1792–1861), meant that others deferred too much. Had Thomas Wilson Croker, the chief critic of Arctic exploration, still been in office this instruction might have been opposed. But he had resigned in 1832 in opposition to the passing of the Reform Bill.

No news was heard for two years and Sir John Ross wrote to the Admiralty Board offering to lead a search expedition. Coming from the still-discredited Ross, and because the Board had not begun to consider any possibility that he was lost, his offer was rejected but forwarded on to Parry with the request that he should seek opinions from Sir Edward Sabine, Dr John Richardson, George Back and James Clark Ross. For unknown reasons Parry did not wait for replies from Sabine and James Ross, but with Richardson's support replied that there was not yet any cause to worry. They did suggest however it would be prudent to *begin* to prepare two separate searching parties along those same lines proposed by Richardson previously. This delay proved fatal.

In the early summer months of 1848 a party of men carrying food and other supplies, led by Richardson and Dr John Rae, set out for the Coppermine River following the same route as Franklin's first land expedition. James Clark Ross also sailed with HMS *Investigator* and HMS *Enterprise* for Lancaster Sound stocked with food sufficient for three years, and ready to overwinter on Somerset Island. From this position sledging parties would be sent south and west and try to link up with Rae and Richardson. Parry was involved with the planning of both expeditions as was Lady Franklin, who tried to persuade him that she should accompany Richardson's group.

He argued that there were better ways of supporting the new rescue operations and together they set out to raise a financial bounty for the whaling ships working in Baffin Bay. This would be awarded on verifiable news about the two missing ships, with Parry acting as one of the referees. The reward started at £1,000 but increased over the next eighteen months to £20,000 (£1.6million).

The searching expeditions were delayed for varying reasons and it was a whaling ship, *The Prince of Wales*, that discovered messages left by HMS *Investigator* in special waterproof cylinders, which informed the Admiralty that no progress had been possible due to heavy ice conditions, and with their food running low they would be returning to England in the spring of 1849. They actually reached London in November 1849 by which time Sir John Richardson, who had travelled along the coast between the MacKenzie and the Coppermine deltas without finding any sign of Franklin, was also on his way home.

Almost four years had passed and the Admiralty, under pressure from Lady Franklin, her friends and the press, now decided to launch new searches. Parry, Barrow and his group of experienced polar officers dubbed the 'Arctic Council' by the newspapers although they never actually met, were consulted again. This 'Council' agreed a plan suggested by Parry which comprised of one search from the Bering Straits eastwards, as it was thought possible that Franklin's route at the end of Lancaster Sound could have ended at Banks Land or down any channel between that and Victoria Island.

Franklin had chosen this route but become trapped by ice in the Peel Channel near to King William Island on the southern flank of Banks Land. This route had been unusually clear of ice when he arrived but as winter came ice floes from the Beaufort Sea sealed it off completely. *Erebus* and *Terror*, even if with double the horsepower they possessed had no chance of escape.

The other search would enter Lancaster Sound, seeking signs of *Erebus* and *Terror* west and north of Somerset Island where James Ross had been forced to give up. Neither searches were successful.

In 1859 a message was discovered – initially written on 24 May 1847 and buried under a stone cairn on King William Island. The first note recorded that all was well, but it was added to by Captains Crozier and James Fitzjames on 25 April 1848. The additions recorded that Sir John Franklin had died on 11 June 1847, and that after another brutal winter both crews had abandoned the ships on 22 April 1848. They had left this message behind after setting out to walk down the coast of Victoria Island to the Coppermine River, where John Richardson and John Rae had been during the summer of 1848.

Some Inuit people saw the sailors as they struggled south but only a few weeks later all of them, excepting one small group that tried to return to the ships, had died, leaving a pathetic trail of discarded equipment and personal possessions along the route.[5]

The Board of the Admiralty devised more plans for searches, some involving five ships working together.[6] Every part of the Canadian polynya was examined either by sea or by sledge, with each part surveyed, maps produced and more research commissioned.

By this time Parry had recognised the inevitable and took little part in the later stages, preferring he said, 'to leave it to those who have succeeded me in the Arctic Regions'. According to his daughter Lucy, he kept a large map in his study of each sledging journey and later organised an exhibition in the Painted Hall at Greenwich Hospital for the personal effects of Franklin and the missing crews that had been found.

There was no acknowledgement that, well-intentioned as it may have been, he had been sadly wrong to support Franklin's application for the leadership of the expedition and that he had failed to give backing to his preferred plan to travel from west to east. He had also been wrong to suggest that search expeditions were unnecessary, until it was too late.

Barrow had been right to favour James Fitzjames, a younger, healthier and more experienced commander who had served under James Clark Ross, but his suggestion for an examination of the Wellington Channel needlessly wasted valuable summer months.

Though Parry had by now lost hope Jane Franklin did not, and continued her demand for more searches. Gradually disenchantment with the search for the doomed Franklin ships and for the costly Northwest Passage project, had set in. The new *Monthly Magazine* in January 1850 was scathing in its criticism:

'A miracle of misdirected energy and enterprise', wrote Francis Ainsworth. 'It is difficult to imagine a Government or Nation seized with the same impulse and communicating it to the crews of so many doomed ships.' The Crimean War news then took over the front pages of most newspapers. John Barrow, who had promoted the project, had died in 1849, and the Admiralty Board from then on expressed far less appetite for continuing his quest, and it drifted out of the public's attention.

Jane Franklin ensured her husband was credited with finding the Northwest Passage although he had not lived to complete it. His route down the Peel Channel was one of several routes and it was by following another channel past the west coast of Boothia and through the Simpson and Dease channel that the Norwegian explorer Roald Amundsen completed it with his yacht *Gjoa* fifty years later. It took him four years to do so.

Chapter 19

The Two Hospitals

Two months after Franklin's departure in May 1845, Parry wrote to George Eden, Lord Auckland (1784–1849) the new First Lord, explaining that he was unable to continue as Comptroller of Steam Machinery as well as head of the Steam Packet Service. There were no deputies to share his work, he was in poor health and physically exhausted.[1] He had had an operation for gallstones in 1845 which was 'attended with complete success',[2] and afterwards produced his final report with proposals for the reorganisation of the Steam Department. Three years later, in 1849, his post was abolished and the Steam Department was placed under a fully qualified naval engineer, the Admiralty at last recognising that the Royal Navy of the future would be powered by steam.

A week after his letter was despatched he was offered the position of Captain Superintendent of Haslar Hospital at Gosport, a five-year tenure. Yet again it was combined with an extra responsibility, this time the Superintendency of the Royal Clarence Victualling Yard at Portsmouth. Neither of these were sinecures but they offered him an opportunity to move out of London to a healthier climate and provided income which was, he said, 'the same as they had been previously'. He wrote about this news to Lady Maria adding:

> the climate is one of the healthiest in England, the duties nothing (to me!) and my time quite at my own command. There is a good garden. I have two cows, plenty of hay and a boat's crew always at my command. Nobody to interfere with me except her Majesty when she passes en route to the Isle of Wight. On which occasions I must attend her embarkation.[3]

The hospital was near Alverstoke overlooking the Solent and nowadays only five minutes from Portsmouth by motor ferry. The Captain Superintendent's house was in a terrace of fine houses and he was given use of one adjoining property because of his large family. From their upper floors it was possible to see out over the sea to the Isle of Wight.

Haslar was a very large hospital accommodating 1,000 patients. It had been built in the 1740s and was the largest brick building in Europe at that time. Run by the Royal Navy it was organised on the same command structure and rules as a man-of-war. Just before he arrived the medical staff had been in dispute with his predecessor who, like him, was a naval officer rather than a qualified medical man. Parry's family background as son of Dr Caleb Parry the eminent physician, helped him overcome any reservations they harboured.

He was always tactful, genial and earned the respect of the staff when he consulted them about long overdue reforms to their working arrangements and conditions. By chance and good luck the Chief Medical Officer was Sir John Richardson FRS, a Scot with, many said, an iron resolution, and who was respected and admired by Parry. He had accompanied Franklin on two expeditions and took part in the subsequent search expedition of 1849–1850.[4]

Both wished to modernise the hospital, intending that it should be more efficient and far more welcoming to its patients. With the help of Catherine, Parry started Bible lessons for the convalescents, and others for those sailors mentally disturbed from the after effects of cannonfire below decks..

The Parry's kept an open house for the officers based at the Royal Clarence and Portsmouth dockyard and encouraged young cadets from the Portsmouth Naval College to visit them for meals, social events and Sunday evening Bible classes.

One of these young officers, Commander Robert Coote, became engaged to Lucy, whose health had become more robust and she had fewer fits. Another daughter, Priscilla, one of his two step daughters, became engaged to Edward Hardcastle, a liaison that delighted both sets of parents. These two betrothals were announced in the spring of 1850 in letters sent to Lady Maria: 'Lucy and he have known each other for some time past, he having been at the Royal Naval College at Portsmouth and frequently visiting us at Haslar. He is a very shy and diffident man (such as you dear Mother, will not, in that respect at all approve!)' This directness in making such a comment indicates the change in their previously fragile relationship. There was, however, one more serious dispute between them still to arise, over Edward's infatuation with his first cousin Maud Adeane, who had been diagnosed with tuberculosis. Lady Maria saw this connection as 'unfavourable'.

It was a happy and settled time for everyone.[5] Parry was delighted with the return of his favourite son Charlie on leave from the Pacific in the autumn of 1850. All too soon, in January 1851, Charlie was commissioned on the frigate HMS *Amphrite* and their brief reunion ended. Parry went to see his son leave Portsmouth and felt sure, as he later wrote to Catherine, he would never see him again. His deep attachment to Charlie is evident. 'Our precious Charlie's leaving home was a really trying thing. I really did not know how to stand it myself.'[6] He saw himself joining his first ship as a volunteer First Class working hard to become a midshipman and well remembered the hardship and trials that awaited his son. When the *Amphrite* sailed away he confessed in his letter that he 'knelt down on that black stone seat and poured out my whole heart, with strong crying and tears on his behalf'. This was a side of Parry's character that few people knew.

Despite his work in the hospital, which he visited daily talking with patients and listening to the staff, he became involved with the Parish of Gosport without upsetting the incumbent priest, and helped found the Portsmouth Sailors Home in 1853. His intention was to improve retirement conditions and welfare for ordinary seaman, and to commemorate the opening of their home he gave a talk to 600 sailors at the Victoria Rooms Southampton on 19 December 1853.

The theme of his speech was his conviction about Christian values and their relevance, and how they were especially relevant for seamen who, 'by spending a very large part of their lives exposed to constant hardships and danger, reckless of life, deprived of the humanising and softening influences of social and especially domestic life, became a sadly neglected class'. He addressed these views directly to the seaman whom 'he thought he knew well', and gave them his advice that:

> sailors had to believe the Sabbath was made for *them*. Officers must act up to the admirable spirit of the first Article of War for the governance of the British Navy. Discountenance everything that tends to the derogation of God's honour and the corruption of good manners. Character and condition are closely connected.

Later in this speech he referred to the fact that many naval officers were still convinced that, 'the character of the British seaman would be lowered in the eyes of the world by any attempts to improve their moral and social conditions on shore', while he made it his constant aim to prove that 'a Christian sailor was not only a better and happier man, but a better seaman also'.[7]

He continued this subject by extolling the work of the Naval and Military Bible Society instigated in 1804, the British and Foreign Sailors Society founded in 1833, and the comparatively new Seaman's Charter (the Merchant Seaman's Act of 1835) initiated by Sir James Graham, First Lord of the Admiralty. The Act set down rules for 'Articles of Agreement and Discharge' from ships to be supervised by its officers, a fund for sailors widows and children, the seamen's Hospital Society and minimum standards for food provided on merchant ships. The expansion of the Royal National Lifeboat Institution, originally founded in 1824, was included as well.

Each of these were important improvements to the dreadful conditions sailors normally endured and which he seen at first hand during his early years in the Navy. These producing 'a degraded class of men given to reckless folly and childish extravagance'. One expression of which was the crude description that life on the lower deck consisted of, 'rum, bum and the Catherine wheel', the latter being the wooden frame to which men were strapped prior to being flogged.

In May 1852 he was promoted to Rear Admiral of the Blue. This came unexpectedly after many years as a captain. The death of more senior officers and his place near the top of the Navy list gave him his 'flag'. This senior rank had one serious disadvantage. He was now considered too senior for his post at Haslar and would have to resign before his five-year term was finished. He would be placed on Admiral's half pay from 4 June and had to make way for his successor immediately.

Their time at Gosport ended abruptly much to Catherine's regret. She had thrown herself into several local charities, one being a Ladies Association for the making of clothes out of discarded clothing from the Naval dockyard for the starving people of Ireland.

Their own family, the patients and the staff at the hospital were all reluctant for them to go and he wrote to his son Edward now a student at Oxford, about their last day at the hospital. He describes their leaving with genuine emotion, 'never, I believe did people part with more sincere regret, and you will not wonder that we felt it a relief when at length the train was in motion for Fareham'. The family left Haslar on 29 July, seen off by the Richardson family, and went straight to Basing Park in Hampshire to stay with his sister Maria and her husband Thomas, Rector of Wickham later to become Dean of Winchester.

They remained only a few days and then visited their step-children and children, including Edward in Oxford and Priscilla and Edward Hardcastle in Keswick in Cumbria, where he hoisted his Admiral's Flag for the first time on a Derwentwater launch. There were expeditions from Keswick visiting Borrowdale, Buttermere, Wastwater, Grasmere – where they met Dorothy Wordsworth – and to Ullswater. Parry delighted in the wild, untamed landscape and 'completeness'. Described by Catherine in a letter to her brother in September 1852. Finally they visited Camefield near Manchester.[8] But his illness returned and decisions were needed over where to find a suitable home.

They found a property called Northbrook House in Bishop's Waltham and moved in December 1852. Once again they took on the work of visiting the poor and ministering to the sick, including their son Willie, who nearly died from what Catherine describes as a 'head attack', possibly severe migraine.

During this period the extended search for Franklin and his missing ships came to a conclusion. Among the various searching parties looking for Franklin and his men was an expedition using the same ships used by James Clark Ross earlier. The *Investigator* and *Enterprise*, who went through the Bering Straits in 1850. *Investigator*, commanded by Captain Robert McClure, rounded Point Barrow and reached the south east coast of Bank's Isand. Here she became locked into the ice and remained for two winters, one of which produced the coldest temperature yet recorded in the Arctic at -55C° below freezing. With food running out McClure decided to take twenty-six men with sledges and trek eastwards where they had seen high land.

After sixty miles of hauling over broken ice floes and large pressure ridges, they reached Melville Island and Winter Harbour, where *Hecla* and *Griper* spent the winter of 1819–20. McClure, his First Lieutenant, Gurney Cresswell, and his exhausted men were then discovered by a searching party from HMS *Resolute*. The *Resolute* picked up his weakest sailors from *Investigator* and prepared to take them home.

McClure, wishing to claim a £10,000 (£680,000) reward for completing the passage, wanted to remain and force his way across in the spring, but Kellett, captain of *Resolute* and his senior in rank, overruled him. McClure was half mad by this time.[9] By travelling over from Banks Land to Melville Island, this link in the Northwest Passage had now been now been made.

Cresswell, son of one of Parry's oldest friends, brought back this news and went to Northbrook to give Parry the details and an invitation to be guest of honour at a special celebration in Kings Lynn where the Gurney-Cresswell family lived.[10] A passage had been found using much of the route Parry pioneered thirty years before. As the guest of honour as well as a Freeman of the City of Lynn, he gave a speech in which he spoke from his heart.

> I came 200 miles to be here and would willingly have come 2,000 to be present this day! How little I thought when I stood upon the western shore of Melville Island and discovered Banks' Island in the distance that, in the course of time, there would come another ship, the other way to meet me and to be anchored in the Bay of Mercy. But while we are rejoicing over the return of our friend and awaiting the triumph that is waiting his companions, we cannot but turn to that which is not a matter of rejoicing but rather of deep sorrow and regret, that there has not been found a single token of our dear long lost Franklin and his companions. My dear friend Franklin was sixty years old when he left this country and I shall never forget the zeal, the almost youthful enthusiasm with which he entered on that expedition. In the whole course of my life I have never known a man like Franklin who combined so many remarkable and different qualities. With all the tenderness of heart of a simple child, there was all the greatness and magnanimity of a hero. When I read of the way in which the last link of the Northwest Passage has been discovered, – it rekindles in my bosom all the ardour of enterprise, aye, and much of the vigour of youth![11]

New proof of Franklin's death came from several sources, including Dr John Rae of the Hudson's Bay Company in 1854, and Leopold McClintock of the *Investigator* in 1859. The *Times* now summed up the mood of the nation: 'We have had quite enough of great Arctic Expeditions; since Sir Edward Parry's first voyage in 1819–20, with single exception of Captain McClure's, they have invariably resulted in disappointment and disaster.'[12]

Writing his account of celebrations at Kings Lynn Parry used extracts from Cresswell's journal and mentions letters just received from Lord Aberdeen and Lord Stanley giving their endorsement for his appointment as the next Lieutenant Governor of Greenwich Hospital.[13] This was followed by terrible news from the Pacific, where his beloved son Charlie had suffered an accident to one eye while skylarking on board his ship.

Parry was appointed Lieutenant Governor of Greenwich Hospital in November 1853 at a salary of £800 (£70,000 today) a year. Just before moving into his new home, Archdeacon Wigram of Southampton invited him to address all the seamen then present in the docks. This lecture was attended by a huge crowd and eventually published by the Admiralty with orders that it was to be placed in libraries on all naval ships. Using terms and language familiar to the common seaman, his speech of thanks expressed what everyone present felt, 'that we ought to be the better for what we had heard'. It was the last public speech he was capable of making.

In January 1854 they were living in the old Governor's residence at Greenwich. The hospital then was supported by a levy of 6d a month deducted from every seaman's wage paid in Britain and in her Colonies.

Unlike Haslar the house had no garden but he approved of the handsome and spacious building and its surroundings, especially Blackheath where James Clark Ross now lived and the open green space of Greenwich Park. 'I doubt not that we shall become very proud of it,' wrote Catherine to her father.

Family reunions at Greenwich were frequent. Captain Coote, Lucy's fiancé, returned from the Niger expedition.[14] Edward, his eldest son, was ordained to Holy Orders in London and became Chaplain to University College Durham, and Caroline Martineau also came to stay for an extended holiday. A severe outbreak of cholera swept through Deptford and West Greenwich carrying off many of the elderly residents of the hospital but the younger patients were kept in isolation.

Parry's stomach pains now became more severe and Catherine wrote that this was an internal disease 'probably of an organic nature', suggesting it was probably some type of cancer. A doctor called Bence Jones was consulted but was unable to prescribe anything new or effective. Sir John Liddell, the chief medical adviser at Greenwich, recommended they visit Ems in Nassau, Germany, to take the mineral waters and consult a German doctor called Soest, who could provide special treatment. Plans were made to leave London on a steamer for Rotterdam on 23 May and a large party accompanied the patient including sister Catherine, the young twins Kate and Bessie, Eliza the cook, Manton the butler and Miss Smith the tutor.

The journey proved to be arduous and he arrived very weak and exhausted. To make matters worse Dr Soest forbade the use of opiates, which had sustained him over recent months. The doctor was sure that he would feel an improvement after the first course of mineral water but this optimism was unfounded. Catherine then wrote affectingly to her father that she feared the worst and described sitting alone in the dining room at their hotel praying silently to God to deliver him from his pain.[15]

The move to Ems was a bad mistake and he became depressed, which the doctor insensitively blamed on his withdrawal from his opiates. The new treatment, Soest believed, would begin to show results in a few more days but his optimism was again mistaken. More help for Catherine was asked for and her father and Joseph Hoare and his wife arrived on 20 June. Another close friend, John Pattison, stayed a short time to relieve Catherine from 8.00pm until midnight, so that she was able to get some sleep. Manton, their manservant, took charge during the rest of the night.

Over the next two weeks letters to friends and relations show the depths of Catherine's despair over his increasing pain and physical weakness. She was at least confident that his strong Christian belief would ensure his joyful reception by God and she describes this time as a personal conflict between herself and her God, not wishing to lose her beloved husband while knowing that he would soon be in the place for which he had prepared himself.

This conviction was explained fully in many of her letters. Eventually the strain became hard to bear and the older children were sent for. Edward, and Caroline arrived at Ems at the end of June and in the first week of July Captain Coote and Lucy also. On 7 July, surrounded by his family, Parry was given the Sacrament by his brother-in-law Edward Hankinson; early on the morning of 8 July, he died.

Arrangements for his body to be returned to England were made rapidly but the journey to Rotterdam by special boat from Koblenz took four days. The family, now thirteen in total, made their way back to Greenwich by steamer while Manton travelled with the coffin. They arrived in Greenwich on 13 July.

His funeral arrangements included a special service held in the Chapel at Greenwich Hospital followed by burial in the Mausoleum. Edward Hankinson, writing to his elderly father, said: 'yesterday passed off very much as you would suppose. It was a large funeral and well arranged.' The procession left the

Governor's House soon after 12pm. In front of the coffin, on which lay his cocked hat and sword, marched a regiment of pensioners with furled flags and creped drums.

As soon as we left the house a large number of officers connected with the hospital fell in from the council room and Arctic officers not a few, Austin, McClure, Collinson, Kellett etc. His coffin rests beside that of the late Governor, Sir Charles Adam and over that of Sir Robert Stopford. This terminates the sad story. A mighty man of valour and a faithful servant of the Lord Jesus whose light has shone brightly before men, and whose record is on high, now consigned to his last resting place among the honoured of the earth and there awaits the morning of the bright and glorious resurrection.[16]

Chapter 20

Epitaphs

Rear Admiral Sir William Edward Parry was buried in Greenwich Royal Hospital graveyard but the coffin was relocated when the new nurses home was built in 1881. During the removal his name and those of several other distinguished occupants, were regrettably detached and then lost. During the Second World War a Luftwaffe bomb fell on the enclosure where his coffin lay, destroying memorials and scattering remains over a wide area.

Sir John Franklin's grave has never been located. It is somewhere on the icy windswept north western shores of King William Island, most probably close to the sunken remains of his ships. In a note written by Captain James Fitzjames found inside a message bottle, Franklin died on 11 June 1847 after two winters in the ice. During the first winter of 1845–46, three men had died and were buried on Beechey Island. They had already come 2,048 miles with approximately 610 miles remaining to be completed, mainly around the Great Bay at the end of the Victoria Strait.

Erebus and *Terror* were abandoned in Queen Victoria Strait in April 1848 at latitude 69 deg. 37' and 42," longitude 98 deg. 41', eighty miles from Cape Herschel on the southern tip of King William Island. Here a large cairn had been erected by two Hudson Bay men, Thomas Simpson and Peter Warren Dease in 1839. In Simpson's estimation Cape Herschel was the hinge for the last section of the Northwest Passage. As Franklin's grave has never been found, this site would feel very suitable as a memorial.[1]

Two large memorials however were built in Britain, one in Westminster Abbey carrying lines composed by the then poet laureate, Alfred Lord Tennyson;[2] The other in Waterloo Place close to the Royal Society, paid for by Parliament after much lobbying by Lady Jane and was erected in 1866. It carries the name of each man who died and this tribute to Franklin:

'To the great Arctic explorer and his brave companions who sacrificed their lives in completing the discovery of the North West Passage AD 1847–48.'

Parliament had been persuaded by Jane Franklin to credit her husband with the discovery of the passage although this was not strictly true. He had got nearer than most, except for Parry, but the attribute was very likely taken to convince the public in general that his personal sacrifice and the huge public expense for the numerous searches, was justified.

Parry has neither memorial nor tribute; only a small plaque on 27, The Crescent Bath that reads: 'Here lived Dr Caleb Hillier Parry b.1755 d.1822 and Admiral Sir W.E. Parry b.1790 d.1855.' No memorial was ever rebuilt at Greenwich and his mortal remains are lost. The British public have nothing to educate or inform them of his pioneering voyages, his family life and varied career. If Parry had chosen his own epitaph it would most probably have been a suitable text from the Bible praising the courage of his officers and sailors. One suggestion was made by his son Edward, that his memorial should carry one of his favourite axioms: 'Man proposes and God disposes'. The exact words used by Sir Edwin Landseer RA for his famous Arctic painting of 1864.[3]

Parry's faith had been severely tested when Isabella died young, and the deaths of five of their eight children. These were tragedies that he quietly accepted, their loss being proof of 'God's will and innumerable mercies', and that however short, 'life was an education for eternity'.

He held to his convictions and continued to teach sailors in both naval hospitals as well as the Australian convicts how to live productive Christian lives. His evangelical work was based on his belief that the advantages for the Royal Navy especially would result in better conduct, self discipline, physical and mental health and lead to an improved standard of new recruits as well as the retention of the best able seamen.

His officers were mostly those who shared his ideals. William Hooper, Edward Fisher, John Irving and William Hoppner. They too had drawn from and seen in the crews that the Polar regions brought 'religious thinking to the surface even in men not much given to talking of such things'.[4] The routine of naval life, harsh though it was, could be improved. Men who had never had proper spiritual guidance became optimistic about their future, joined in with literacy and numeracy classes and willingly took part in church services.. The same men chosen primarily for hardiness, seamanship and self-discipline usually re-enlisted with Parry who knew that this transformation increased their reliability and determination despite the very worst privations they underwent.[5]

With Isabella's support in New South Wales he followed the same principles among the convicts and early settlers placing emphasis on trust, honesty and the need for truthful and peaceful relationships. In many cases he found the convicts more amenable to his teaching than the settlers.

His second marriage to Catherine Hoare, with her family's clerical background, strengthened his faith despite the tragic loss of Willie – his son with Catherine – from a chill contracted at Harrow School. Some historians accuse Parry of being overly pious but his faith carried him through life's tragedies and disappointments. His musical and theatrical talents inherited from his mother Sarah, gave much amusement to sailors during voyages, and he looked for opportunities to hold special celebratory galas and Christmas feasts on board. They entertained many young officers and their friends at both naval hospitals, and joined in with charades and plays. Edward wrote, 'few have ever exhibited a more striking refutation of the charge, often bought against religion, of a tendency to cast a shade of gloom over the pleasures of life; his piety was as cheerful and genial as it was active and practical.'[6]

He both wrote and performed plays on *Hecla* and took part by dressing up as a woman in several, which, under the strict rules of conduct laid down for naval officers, was unconventional, even foolhardy. The sailors liked his approachability and his interest in all matters of welfare on their behalf. They became members of a family dependent on each other, having complete trust in their captain.

Franklin was a devout and strict Sabbatarian. His first wife Eleanor Porden protested at some of constraints he imposed to protect the day of rest. He even showed reluctance to allow her to follow her own interests such as verse writing. During expeditions he chose to recite poetry by Milton and Dante during marches across the desolate tundra and during the long winters cooped up in Forts Enterprise, Franklin and Resolution. Only one entertainment was permitted, an ingenious puppet show organised by Lieutenant George Back.

Although amiable and charming, Franklin could seem aloof due to his increasing deafness. He was respected and liked by his men but his stern principles precluded activities which could create the friendly relationship of 'family' that Parry aimed for. During his years on the Channel Blockade, Parry saw that physical exercise, music and other distractions, were effective ways to avoid depression and mental problems. His use of plays using as many individuals as possible to keep them occupied and working together was for this reason. Ships gossip and light-hearted stories were included in the weekly newspaper Sabine produced and he led communal singing around the barrel organ, allowed dancing and organised musical evenings for his officers when he played his own violin. All were means to the same end and few cases of drunkenness, insubordination or aggressive behaviour were recorded in the log. Flogging was never practised.

Parry and Franklin felt certain that flogging brutalised its victims as well as its perpetrators and was contrary to their shared conviction that seamen should be treated decently and well.[7] Franklin's wartime record was impressive, having been with Admiral Nelson during the famous battles of Copenhagen and Trafalgar, and part of the amphibious attack on New Orleans in 1812. He, unlike Parry, might have viewed his expeditions as an extension of war – now against the ice rather than the French – but as voyage accounts were largely written by others and he kept no diary, it is impossible to verify this assertion either way.

The years of navigational experience in Arctic waters which Parry had on 'special operations' in the Baltic, North America and Norway were a sound background for his later years. His interests, including geology, astronomy, botany and meteorology, came from Caleb and he became a good observational scientist which put him on equal terms with astronomers, surveyors and geologists.[8]

One reason for his success was the personal attention he paid to checking and evaluating every item taken on board even minor details received personal attention, whether rigging, heating, bedding, food, medical supplies, clothing, or different types of insulating materials. He paid close attention to hygiene and ventilation in the cramped mess decks and adapted some practices to suit the extreme Arctic weather.[10] He also designed a portable observatory strong and sufficiently weatherproof to be used all year round, but still light enough to be carried with only a few men away from the magnetic field created on every ship by its metal fittings.

He was a man of physical energy, always preferring to be doing things however arduous or dangerous rather than delegating to others. The long trek in late winter of 1820 on Melville Island, surveying sections of coastline in small boats in freezing and foggy weather in Hudson's Bay, and weeks of sledging across the Melville Peninsula in 1823 and 1824 all contributed to his rheumatism, while the march to the North Pole worsened it so much that he could barely continue.

In New South Wales he travelled by horse or light carriage 'seeing for himself', spending weeks away from home inspecting farms, sheep stations, land grant options as well as the Newcastle coal mining business. Once more in England he was to travel throughout rural Norfolk and Suffolk on roads and lanes indifferently maintained by the parishes and almost impassable in winter. Because the Poor Law Commissioners needed their new regulations imposed thoroughly and consistently, his evenings were often then spent writing reports on progress and the many objections he had encountered. His one consolation being that Isabella and the children now had a safe and quiet home far away from the cholera epidemics periodically raging through London. A short period of convalescence would have helped, but he was appointed to organise the Postal Packet Service and forced by this position to travel long distances in all weathers.

He was exploited by the Admiralty, who usually gave him two roles and paid for only one. They invariably quibbled over the staffing he needed which limited his attainment of goals that he had identified and knew were achievable. This angered and depressed him, but despite all this his sense of duty never diminished. In letters to Charles he admitted that his income was a constant concern and that he had to accept any new commission offered.[11]

His brother had inherited most of the family properties in Bath, including Summerhill. Charles had enough income and time to transcribe and publish the medical papers their father had prepared.[12] But Caleb had left his financial affairs in a confused state, with money owed by unreliable tenants and other individuals. As a consequence they were both drawn into long-lasting lawsuits. Summerhill was not an easy house to sell and had to be left unoccupied while the ground rents from several family properties in Sion Place were being disputed.

His young children, including Lucy the diabetic, needed special tutoring and medical attention, while Isabella never enjoyed robust health and needed frequent medical advice during her many pregnancies.

Just as he returned to full, as opposed to half pay, other problems arose such as the collapse of the Bank of New South Wales; London needed to be their home and the cost of living was high. His second wife, Catherine Hoare, had her own income which strengthened their financial position, and his postings to Haslar and Greenwich Hospitals with accommodation included reduced the strain, but the benefits came at a late stage and were only a minor compensation for those long years during which his pay bore little relationship to his responsibilities.

He confided in Charles and Caroline throughout his life and usually only with his other two unmarried sisters, Matilda and Gertrude, about their mother. Ten years absence on expeditions followed by four years in Australia created a wide gap between himself and his nephews and nieces. He was a reserved man and his letters contain little about his relationships with his Stanley in-laws except for Lady Maria, whom, strangely he became much closer to after Isabella died!

All his sisters except Caroline predeceased him. Gertrude Trevor in 1848; Maria, the wife of Thomas Garnier, in 1849; and Matilda in 1852. In one long letter to Charles after his landmark voyage of 1819–20, he describes his heartfelt desire to 'belong somewhere'. His pressing need for income meant that only the four years at Tahlee and at Haslar Hospital represented a permanent home.

Isabella and Catherine were both extraordinarily selfless women and contributed hugely to his career and personal happiness. Isabella's weak constitution was overcome eventually by frequent pregnancies but she still committed time and energy on her charitable work among the poor. In New South Wales she was the teacher of Bible classes at the new primary school and the small church at Tahlee. She had a loving and quiet temperament, supporting Parry in his evangelical work and rarely expressed a complaint. Parry wrote in his work *Thoughts on the Parental Character of God*: 'the chief comforter of my earthly pilgrimage – the sharer of every joy, and the alleviator of every sorrow – but a faithful counsellor and friend, through many a rough and thorny path in our journey'. She also demonstrated an extraordinary degree of courage during her many illnesses and difficult pregnancies.

Catherine, whom he had known for some years, stepped into her shoes admirably. Her adoption of Isabella's children came naturally, as did the same willingness to work alongside him in both hospitals. Her influence persuaded him to write his pamphlet, 'the Parental Nature of God', and the publication of a 'proposed Union in Prayer for the Promotion of Religion in Her Majesty's Navy'.[13]

Barrow praised most of his Arctic commanders, with the exception of John Ross, for their 'persevering and adventurous spirit in unknown frozen seas making new discoveries in geography and in objects of scientific research, experiments in meteorology, natural history and in short by extending the limits of human knowledge'.[14] It appears now to be a self-justifying statement since much of the information gathered at great cost and Franklin's life was of little immediate use. The cost of the Franklin searches – of which there were thirty in total by land and by sea – have been calculated at £675,000 (£54 million today), and those previous voyages, including Parry's, Franklin's, George Back's and Frederick Beechey's, spread over twenty-five years, a similar figure.

Quantities of data, much of which was duplicated, was stored by the Royal Society, the Admiralty, and the Royal Geographical Society. When a new expedition to the North Pole led by George Nares with HMS *Alert* and *Discovery* started between 1875–76, the same organisations provided manuals which by then had content in depth but which failed to combine the knowledge gained concerning travel, footwear and other practical necessities for travel over the ice.

The same mistakes were made with crews again expected to take an extensive range of measurements and readings, including studies of meteoric dust, using specialist instruments. This work required a team of trained scientists but the same excuse about lack of resources came from the Admiralty, which blocked any such approach.[15] Nares eventually reached 83 deg. north 20' using sledges, but was still over 150 miles from the Pole.

Parry's contribution to Arctic exploration and science was to pioneer means by which ships and men could survive for twelve months or more in sub-zero darkness. He proved that by using the right equipment, well-balanced provisions and diversions to maintain good morale and physical condition in his crew, the exploration season could be extended by years, not just months. These advantages gave The Royal Society and the Royal Navy a world-leading role in several ocean sciences, including meteorological, geological, physical, hydrographical and biological research later evolved by Britain, France and the United States and known nowadays as oceanography.[16]

Parry explored and mapped swathes of the Canadian Arctic coastline (Map 2). With *Hecla* and *Griper* in 1819–20 he found one route through Lancaster Sound that took them further west than anyone previously, while his surveying and detailed charts for Hudson's and Repulse Bay and the north eastern coast of the Melville Peninsula, completed an extensive area of new territory. He believed that a north-west passage existed, but by supporting this view, and with his personal support for Franklin's appointment as leader of the expedition of 1845, he became closely implicated in the terrible disaster that ensued. His statement given in his journal from the third voyage is the position he had reached after nine long years of endeavour:

> May it still fall to England's lot to accomplish this undertaking, and may she ever continue to take the leading enterprises intended to contribute to the advancement of science and to promote, with her own, the welfare of mankind at large! Such enterprises, so disinterested as well as useful in their object, do honour to the country which undertakes them even when they *fail*: they cannot but excite the admiration and respect of every liberal and cultivated mind: and the page of future history will undoubtedly record them as every way worthy of a powerful, virtuous and an enlightened nation.

It was now the scientific purpose which he considered the most relevant, rather than the discovery of any passage there might be.

Sir Clements Markham in *Lands of Silence* described Parry as 'a very perfect sailor, thoroughly well read in all that concerned his enterprises, thoughtful and level-headed, deeply though inconspicuously religious. The *beau ideal of an Arctic Officer*'.[17] John Barrow's tribute was equally fulsome: 'His merits as an officer and scientific navigator are of the highest order. His talents are not confined to his professional duties but that the resources of his mind are equal to the most arduous situations and fertile in expedients under every circumstance, however difficult, dangerous or unexpected'.[18]

His chief qualities were courage, modesty, patience, loyalty and complete integrity. His weaknesses were his ambition, allied at times to snobbishness and his inherited weakness at managing his personal finances. His life was dedicated to his God, his family, and ways he could improve and enrich the lives of common sailors.

Maps

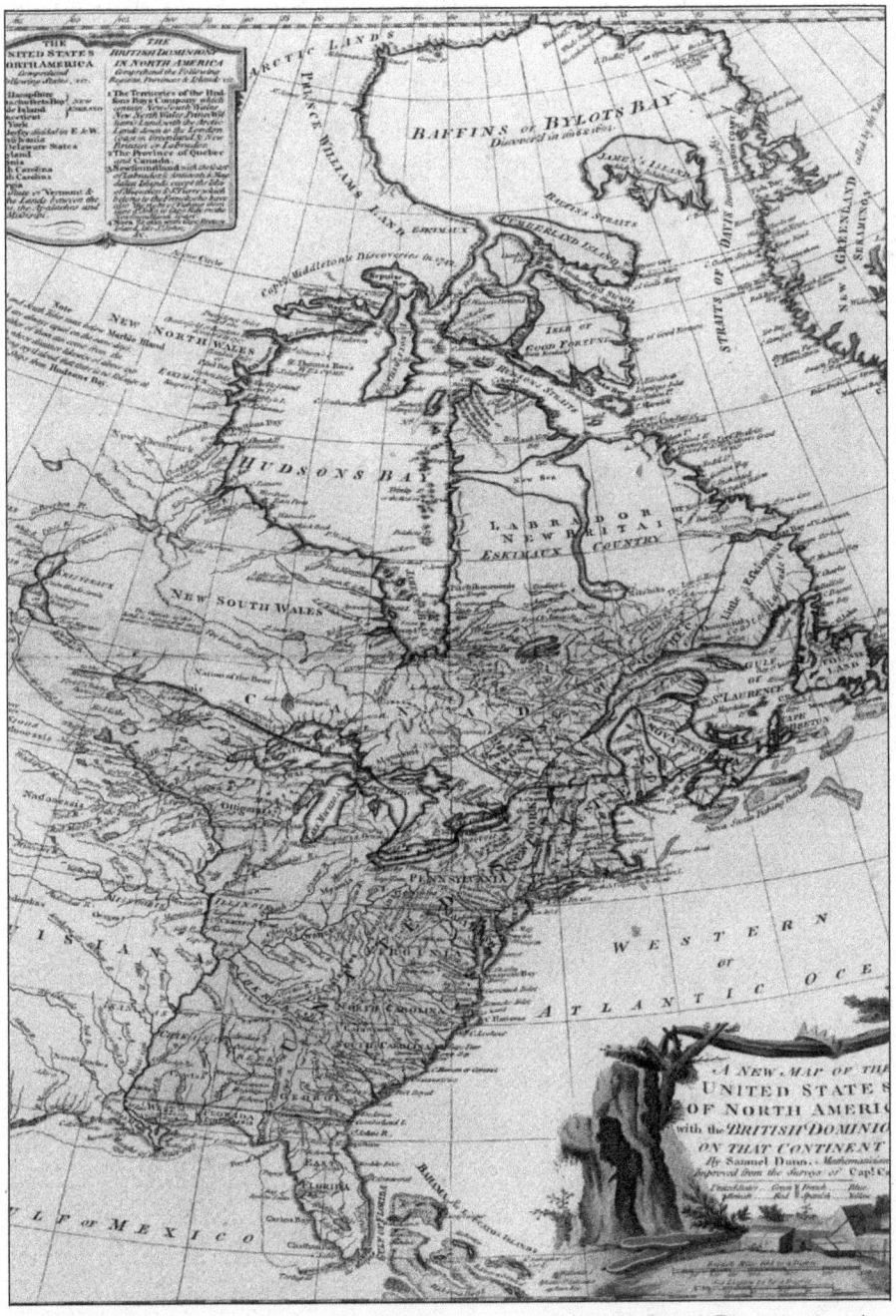

Map 1a: A new map of the United States of North America with the British Dominions on that Continent. S. Dunn from Survey by Captain Carver. 1788.

Map 1b: Polus Arcticus. Septentrionalium Terrarum Descriptio. Gerard Mercator. 1641. Note the Strait of Anian top left.

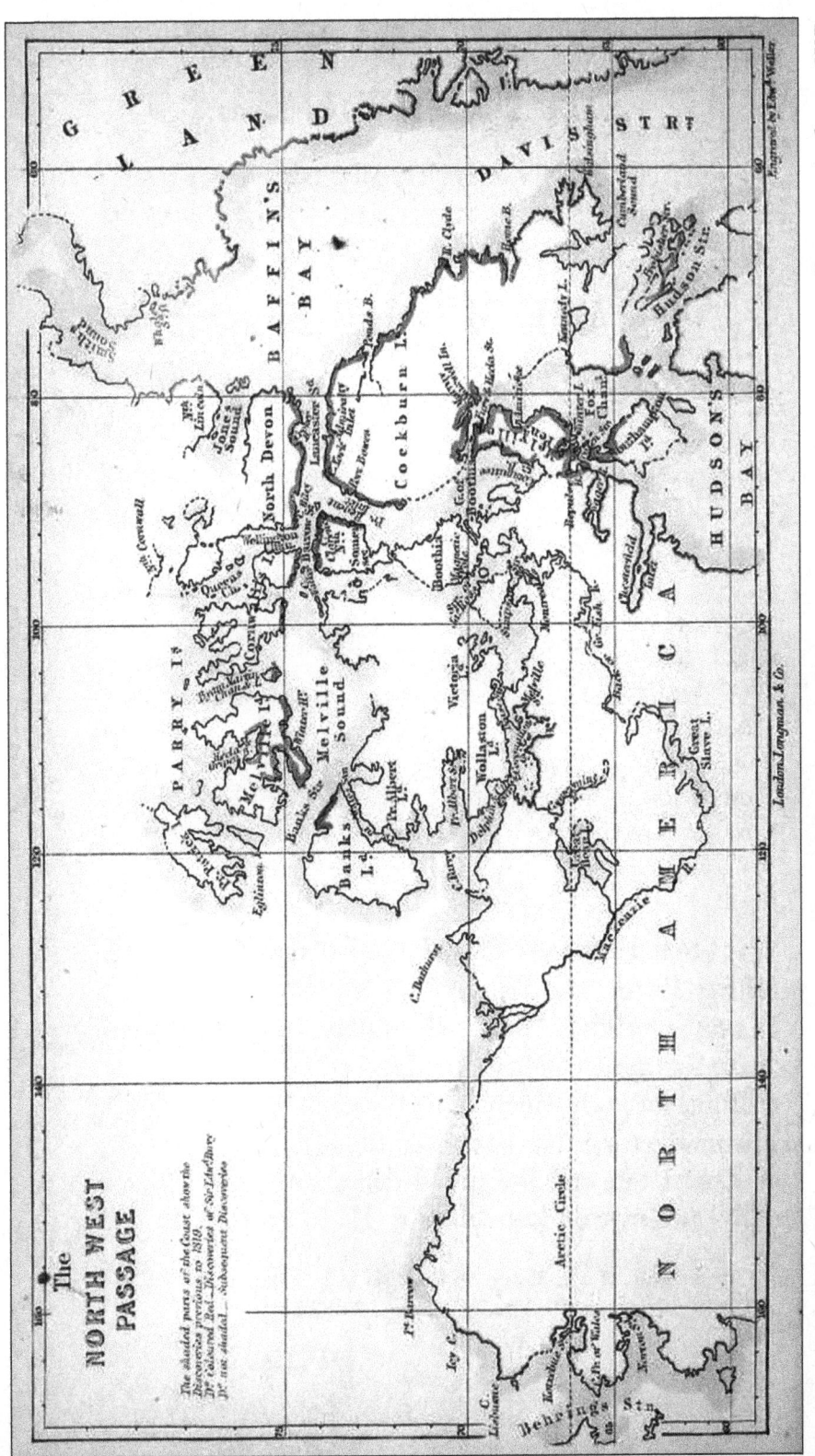

Map 2: The Northwest Passage showing the coast prior to 1819 and the discoveries of Sir Edward Parry and subsequent discoveries. Discoveries by Captain W.E. Parry shaded dark. E. Welley. c.1854.

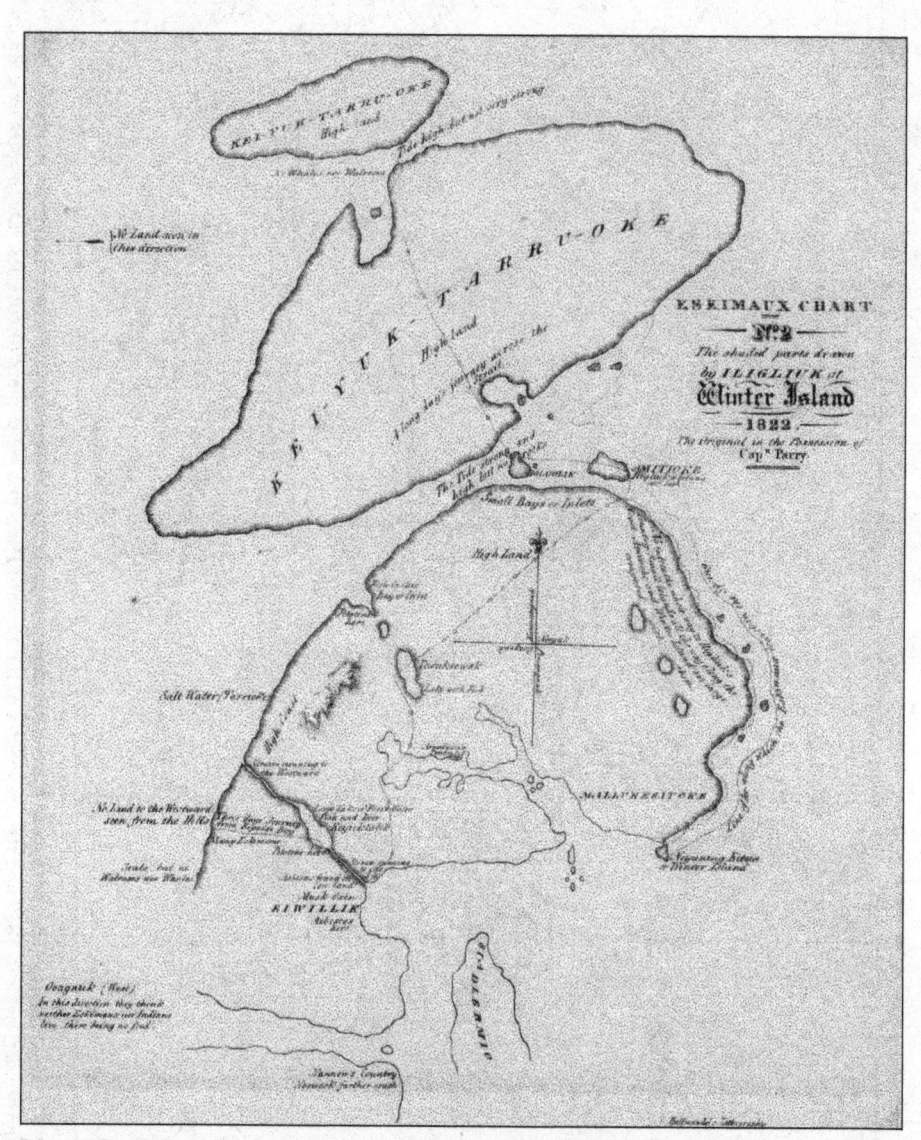

Map 3: Chart No 2. Drawn by Iligliuk at Winter Island 1822. Showing the East and Western coast of the Melville Peninsula and the Hecla and Fury Straits top centre.

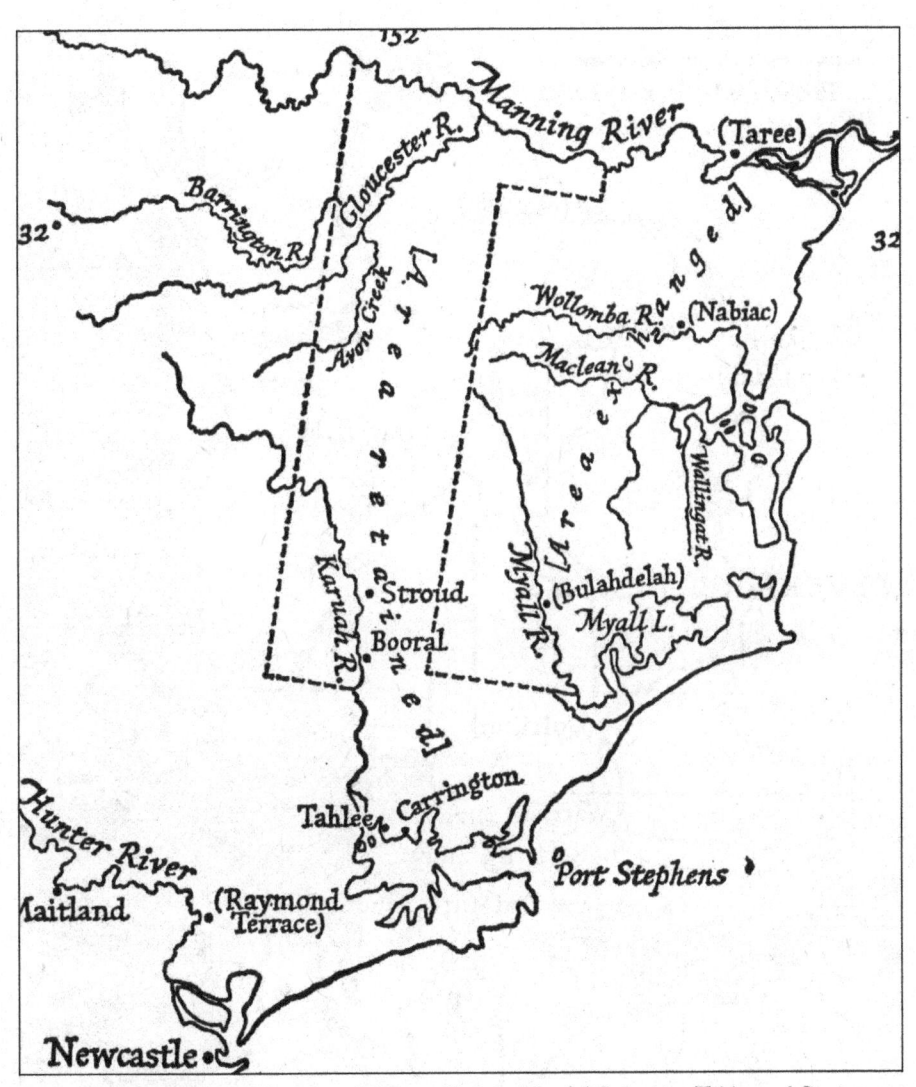

Map 4: Port Stephens and the original Grant of Land of the AAC showing Tahlee and Carrington. S.H. Roberts. 1924.

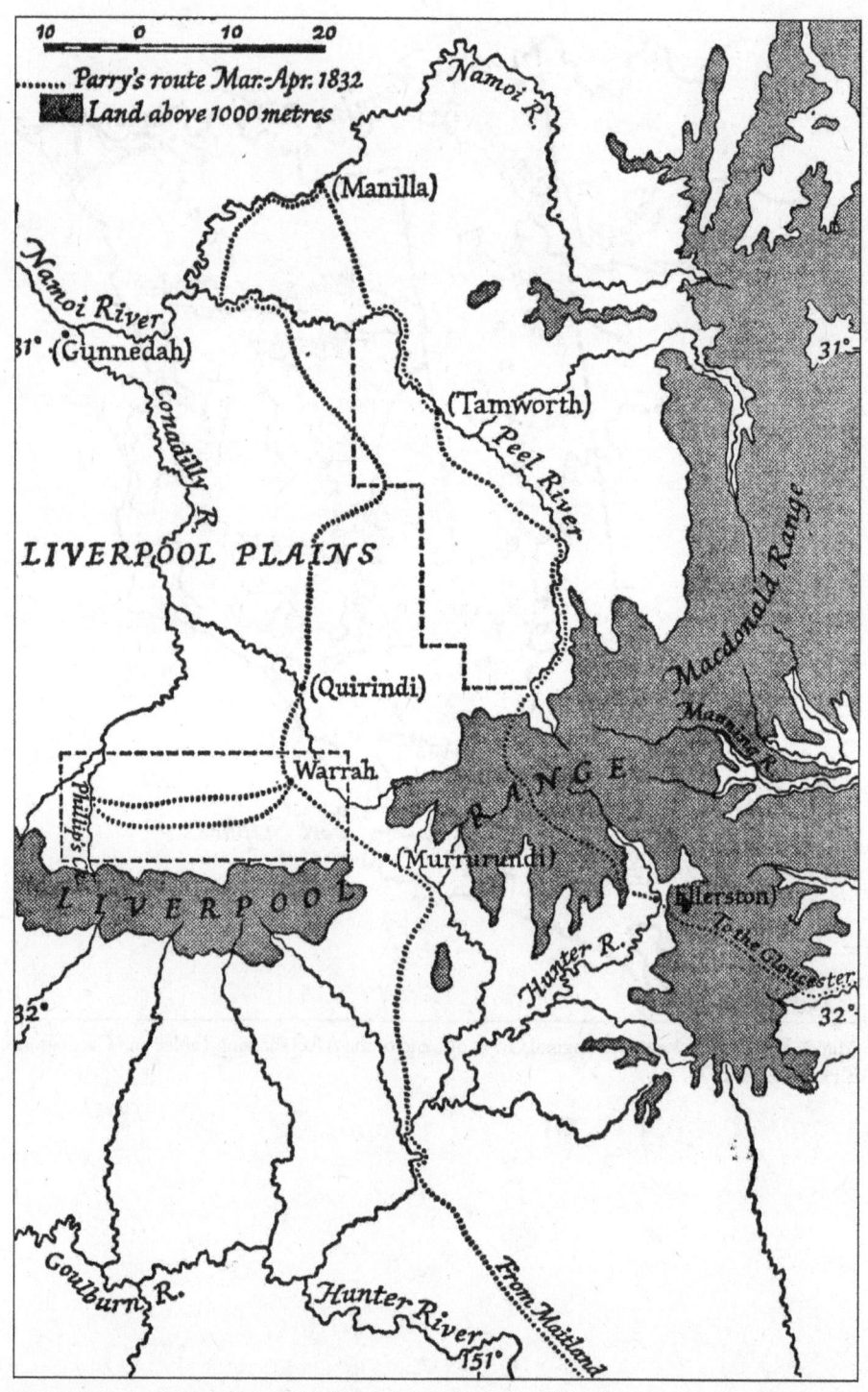

Map 5: Parry's expedition through the Liverpool Range to Quirindi and Manilla and return via the Gloucester River. March–April 1832.

Appendix 1

Geomagnetism

In both the seventeenth and eighteenth centuries studies and records of the earth's magnetic field had been initiated by both the Royal Society and the French Academie Royale des Sciences. Captain Edmond Halley FRS had a special interest in magnetic variation and how it varied over time at any given point. He took his ship, the *Paramore*, on three ambitious voyages during the 1690s sponsored by the Royal Navy and the Royal Society, making 150 observations over the Atlantic. On his return he published an isogonic map which showed lines connecting points of equal intensity of magnetic strength which varied between locations. His map hung in the Royal Society's Meeting Rooms and demonstrated the Society's mastery of matters crucial for navigation. A good source of information on these subjects being *Ingenious Pursuits; Building the Scientific Revolution*. L. Jardine (Abacus Books, St Ives, 1999). Halley's work was pioneering yet incomplete and a thorough study of the phenomena worldwide was proposed. A task that the Royal Society set out to complete in the second decade of the nineteenth century.

On larger vessels and those of the Royal Navy especially, regular measurements were kept and recorded in the ship's navigational logbook. These showed the compass needle deflection from 'true' north for steering, dead reckoning e.g. course setting, and chart correction. Accuracy being crucial for the safety of the ship. The major maritime nations, England, France, Holland, Portugal and Spain compiled this data on a worldwide basis, so that as trade and colonising activity developed so too did the measurement of geomagnetic force in different parts of the world. During the eighteenth and early nineteenth centuries natural philosophers, hydrographers and navigators were searching for possible reasons to make sense of the phenomenon. Many theories and numerous pamphlets and books were written with widely varying magnetic values predicted for each of the main oceans. Some theories were prompted purely by the prospect of financial reward while others were combined with other navigational problems such as the calculation of longitude. Parry, who was absorbed in the studies of magnetism, meteorology and astronomy, wrote a short pamphlet for his sister Caroline explaining the latter while serving on HMS *Alexander*

Parry with Sir Edward Sabine (1788–1883) contributed to the study on magnetism commissioned from the Royal Society during his Arctic voyage of 1819–20. Sabine was awarded the Copley Medal by the Geographical Society in 1821 and was appointed leading surveyor of the first systematic survey Ordinance Survey of the British Isles in 1834 and organised magnetic observatories in British Colonies overseas. See *Observations on days of unusual Magnetic disturbance Made at the British Colonial Observatories under the Department of Ordnance and Admiralty.* (Longmans, London, 1843.)

The readings taken at sea and on land measured both declination and the intensity of field strength. The declination being the horizontal angle between the *local* magnetic and the *geographic* north, or between the compass needle and the local meridian. To measure this, an instrument called an *inclinometer* or *dipping needle* was used (see Fig. 20). This instrument, of which there were several prototypes, consisted of an iron needle pivoted within a circular brass ring. It was mounted on a tripod base with screw feet for keeping itself level. When the magnetic pole came closer the tilt of the needle became vertical, the further away the greater the declination measured.[1]

These variations are now understood, although the composition of the planet has only been known for one century. At its centre solid rock is replaced by molten metal in which there is a gigantic crystallised iron spheroid. The combined core and its enclosing hard mantle are estimated to be 3–6,000° Centigrade. As the earth spins on its axis the immense heat becomes a magnetic dynamo producing electricity. This dynamo, established at the time of the earth's creation, is powerfully amplified and induced currents are transmitted unevenly to its surface.

On the seabed and on parts of the shoreline, bodies of iron ore exist which also affect ships' compasses, and in addition to these the sun also has a strong magnetic force which produces magnetic storms of ionised gas. The same storms produce the aurorae seen in the northern and southern polar regions with charged particles in the storms colliding with oxygen and nitrogen atoms in earth's atmosphere causing colourful flickering lights in the heavens. Parry commented in a letter to brother Charles on 20 January 1826, that, 'the sun and the moon are the causes of the Diurnal variation of the magnetic needle. The sun being the power that produces the *regular* changes and the moon modifying those changes. Extraordinary 'jumps' occur which can only be referred to as electricity'.[2]

At Port Bowen on his third expedition the variation was measured at 6 or 7 deg. and in the letter he names Herschell, Arrago, Barlow, Christie and Babbage, all eminent members of the Royal Society, as being 'the principal individuals engaged in the experiments'. Parry and James Clark Ross, after comparing strengths of magnetic variation with appearance of the Aurora Borealis, concluded that there was no evidence of any connection. They produced a paper on this topic given to the Royal Society on their return in 1825.

Air temperature also affects magnetic influence and ships' logs record temperatures alongside the inclinometer readings as a further guide to the declination expected. Nowadays onboard computers using GPS, the frequent updating of charts and pilot books and sophisticated electronic systems measuring wind and tide give navigators positioning information that is highly accurate, despite the shifting magnetic fields deep below the earth's surface. Geophysicists even today are uncertain how the molten iron geodynamo at the core of the planet moves over time.

Appendix 2

Polar Geology

Various samples of rock, fossils and minerals, with notes describing where they had been found, were collected by Parry and his officers on all four expeditions of 1818, 1819–20, 1821–3 and 1824–6. During the first with John Ross, most were found on land and some as fragments embedded in icebergs from the Greenland glaciers. Others were lifted from the seabed using an ingenious new grab invented by Ross. These were limestone samples which contained a wide range of fossils either being plant material such as tree ferns and small trees, or shells and crustacea of different kinds.

From the three expeditions that followed his personal interest in geology was supported by James Clark Ross, Edward Fisher, William Hooper and Edward Sabine. His midshipmen did sketches of cliff faces and those rock formations that were especially interesting. One such being the sketch made by Midshipman Henry Head in August 1825, showing the immense cliff face behind the *Fury* stranded on the beach below (see Fig. 14). The samples and drawings were sent back first to the Admiralty and then to the Royal Society. The University of Edinburgh also became involved and Dr Jameson, specialist in the new science of geology gave his opinions in notes sent to the Wernerian Society in Edinburgh. Professor Buckland then produced a paper entitled: *Remarks Tending to explain the Geological Theory of the Earth*. The text was printed in *Memoirs of the Wernerian Society Natural History Society for the Year 1826–1831*. Vol IV. Longmans, Rees, Orme, Brown and Green in 1832.

The lecture and Jameson's observations were reported in the *Glasgow Morning Herald* of 27 July 1826 in an article stating that they afforded;

> facts and inferences of a most unusual and interesting kind, primary, secondary and transitional rocks abound. No true or modern volcanic rocks were met with. Alluvial deposits are not extreme. The islands among the secondary traps of Baffin Bay had been filled with secondary limestones, sandstones, gypsum and coal, and above these, tertiary rocks. These had all formed part of the continuous mass with the continental parts of America. That previously to the deposition of the coal formation – as that of Melville Island, the primitive hills and plain supported a rich and

luxurious vegetation principally of cryptogamous plants (algae, lichens, mosses and ferns), especially tree ferns, the prototypes of which are now met with only in the tropic regions of the earth. The fossil content of the secondary limestones also indicates that before, during and after the deposition of coal formation the waters of the ocean were so constituted to support polyp aria closely resembling those of the present equatorial seas. These now frozen regions supported forests of dicotyledonous trees including oaks maples and magnolias, as shown by the dicotyledonous woods met with in connection with these strata in Baffin's Bay, Melville Island, Cape York and Byam Martin Island. That the boulders or rolled blocks met with in different quarters and in tracts distant from their original localities afford evidence of a passage of water across them and at a period subsequent to the deposition of the newest tertiary rocks.

The article continued by explaining that:

black bituminous coal, the coal of the oldest formations which speculators maintained as being confined to temperate and warmer regions, was now shown by its discovery on Melville Island far to the west of Jameson's Land in Old Greenland, to form an interesting and important feature in the geognostical constitution of the Arctic countries. The same general agencies Must have prevailed generally during the formation of the solid Mass of the Earth. It is then indeed that the mind obtains those enduring and sublime views of the power of the Deity.

The report carried weighty evidence that earth had evolved through many millennia and was not created according to Biblical teaching in seven days around 2000 BC. The polar regions had once enjoyed a climate comparable to the tropical regions of the equator in which many of the same plants and animals thrived.

These findings made awkward reading for 'creationists' and can with some confidence be recognised as the initial steps towards the understanding and recognition of an ice age period many thousands of years earlier, and of the ice sheet then covering the globe shaping the landscape features and lakes of the northern hemisphere.

Appendix 3

Arctic Botany

The plant specimens collected and pressed on Melville Island in 1819–20 by Parry, Fisher, Ross, Sabine, Beverley and Edwards were described in the official voyage records held by the Admiralty. Some were given to the Royal Society and others ended up with private collectors. Parry, seemingly less dedicated to this field of research, allowed the dispersal of his own mounted specimens, many of which were given to various friends and institutions. Sabine lent his collection to Robert Brown but asked for their return before he left England in October 1821. There is now no trace of where or to whom they later belonged. It is uncertain whether all their discoveries were fully catalogued since Brown admitted that he had taken too much time to bring the list together due to ill health and a problem determining many of the species. They were, he wrote, of a *very* variable nature and he could find no authentic specimens from elsewhere with which to compare them.

A catalogue of the plants was eventually published under the title, *Chloris Melvilliana. List of Plants collected in Melville Island (Lat 74 -75 Deg.N. Longitude 110 Deg -112 Deg.W.) under the orders of Captain Parry. Robert Brown FRS and L.S.* Clowes, London, 1823. This publication now held in the University of Michigan Library included preparatory drawings (figures) for the engraver prepared by Bauer. Brown made an important comment in his preparatory remarks to the first list of plants. A new genus was also identified and named *Parrya Arctica*: 'In my earliest observations on this subject I had come to the conclusion that from 45 deg. as far as 60 deg. or perhaps 65 deg. of North Latitude, the proportion of *Dicotyledonous* to *Monoocotyldonous* plants gradually diminished.'

But from a subsequent examination of the list of Greenland plants given by Professor Giesecke, an article entitled *Greenland* in Brewster's Edinburgh Encyclopaedia by him stated:

> as well as from what I had been able to collect respecting the vegetation of Alpine regions, I had supposed it not improbable that in still higher latitudes, and at corresponding heights above the level of the sea, the relative number of these two divisions were again inverted; in the Greenland list of plants referred to *Dicotyledones* being to *Monocotyledones* as four to one

or in nearly the equinoctial ratio: and in the vegetation of Spitzbergen, as well as it could be judged of from the materials hitherto collected, the proportion of *Dicotyledones* appearing to be still further increased.

The inversion in the cases now mentioned was found to depend on at least as much on the reduction in the proportion of *Gramineae* as on the increase of certain Dicotyledonous families especially *Saxifageae* and *Cruciiferae*. The flora of Melville Island however, which, as far as relates to the two primary divisions of *Pheanogamous* plants is probably as much to be depended upon as any local catalogue hitherto published, leads to very different conclusions; *Dicotyledones* being in the present list to *Monocotyledones* as five to two, or in as low a ratio as has been anywhere yet observed; while the proportion of grasses, instead of being reduced, is nearly double what has been found in any other part of the world.' See Humboldt in Dictionary of Sciences Nat.tom. 18. p.416. This family forming one fifth of the whole *Pheanogamous* vegetation.

The respected botanist, ecologist and bryologist Dr N.V. Polunin (1909–1997) later wrote papers under the heading 'Parry's Plants'.

A further list of plants and specimens collected by Parry on Melville Peninsula, 1821–3 during the second expedition, was published as an appendix to his journal. This list was reviewed by W.J. Hooker FRS who prepared a paper in 1825 on the botany of the second voyage. This collection contains nine different varieties of saxifrage: Parrya Arctica, Parrya Nudicaulis and a Polygonium Viviparum, and included more specimens from other much later expeditions up to 1887. It was given by Mrs Seton Dickson to the Pollok Morris Collection at the Royal Scottish Museum.

Brown does not go further than to comment on the *Acotylednes* specimens and their high ratio in comparison to others. His opinion was that the abundant game and bird life Parry found during their march over Melville Island in 1820, and the summer camp sites of the Inuit were due to the unusually wide range of vegetation, especially the grasses and sorrel. As summer food for birds like Arctic geese and ruminants so far north it was unexpected and surprising.

The discovery of sorrel growing everywhere, Parry used as an anti-scorbutic, and had the expedition been forced back to Melville Island for another winter it would unquestionably have saved lives.

Appendix 4

Meteorology

As with magnetism, meteorology was another research priority on all Parry's expeditions. Both William Scoresby Jnr. and Bernard O'Reilly completed sketches of many of the curious atmospheric effects resulting from low temperatures and refracted light. Scoresby's work, *Accounts of the Arctic Regions with a History and Description of the Northern Whale Fishery. 3 Volumes,* are a guide to the various kinds of phenomena and went with Parry on all his voyages.

The meticulous voyage records he kept show the positions of his ship each day at noon together with thermometer and barometer readings, the specific gravity of the sea water, wind strength and direction, and the tidal range and velocity. The prevailing weather patterns are also described. Temperatures were taken at many times of the day and night and calculated to find a minimum and maximum mean temperature for each day.

Scoresby's work during his whaling voyages between 1804 and 1817 was the main source of information on polar weather patterns. One limitation was that his tables were based on summer months – April to September – when the whale fishery was open. The predictive value was far less useful for winter months when ice conditions and the movements of high and low pressure systems were altered. But the summer months were those most needed for exploring and surveying safely.

Scoresby had twelve years in which to measure wind strength and direction in latitudes between F.66.59' and F.80.10' deg. N. and from these he extrapolated probable temperatures and ice conditions at the North Pole. This information was important for Parry's attempt to reach the North Pole. Scoresby had refused to support Barrow's conviction about an open sea circling the Pole and was belittled by Barrow as a result. Scoresby's predictions were based on sound observational science and are surprisingly accurate.

The conclusions he reached were:

Swings in atmospheric pressure in the Arctic spring of each year were rapid and portentous. A fall in barometric pressure to 29 inches invariably being followed by a gale, either at the place of observation or very close to it. Both Scoresby and Parry recorded that the oscillations of the thermometer were normally *predicted* by their barometers so that there was 'not an interval of five minutes between the most perfect calm and the most impetuous storm'. Sudden fluctuations in the barometer were indicative of unsettled weather while a fall of more than an inch in twenty-four hours meant a severe and protracted gale. A westerly gale with snow brought about the greatest drop of the mercury and light easterly winds with dry weather, the greatest elevation. Mercury rising high and becoming stationery in early summer meant continuance of fine weather, but in both June or July foggy conditions. Barometers and the accuracy they achieved played a significant role in the management of any ship. Scoresby and Parry both noticed the extreme dryness of the atmosphere and the great intensity of light. During cold conditions this dryness had a remarkable effect on the ship's timbers so that interior panels of wood shrank by as much as half an inch. This dryness was supposed to be the reason for a lack of rust on metal fixtures on deck, their tools and instruments.

From Scoresby's observations, sketches and journals the ice 'blink' was one of the most unusual and distracting phenomena, It was helpful at showing where the ice pack lay.

The 'blink' consisted of a dark or black underpart to clouds, and experienced whaling men could interpret this to find where open water existed in pack ice. It was created by the reflection and refraction of light over ice and very cold water. Objects such as ships seen on the horizon seemed to be lifted above it to as much as four or more minutes of altitude. They also appeared larger as well as elevated and were described often as 'looming', with their lower parts connected to the real horizon by a columnar or perpendicular extension. This optical illusion usually presaged the arrival of an easterly wind.

The *parhelia* and the *corona* also exhibited a wide and varied range of shapes and colour combinations.

Complete circles of light and smaller arcs created halos, with their main sections or *parhelia* all surrounding the sun like some brilliant rainbow; others had their epicentre on its circumference and one spectacular effect was a huge circle of luminous white light coming from the centre of the sun in a direction parallel to the horizon.

Foggy conditions and snow showers magnified the power of these radiant, and extraordinarily beautiful effects which, in Scoresby's own words 'were productive of sensations of admiration and delight'. Parry had one *parhelia* named after him a century later.

It has only been recorded six times since Parry and Sabine first described it (see image below). (C.S, Hastings. *Monthly Weather Review*. Washington 1920.)

New descriptions and records of the weather patterns as far north and west as Melville Island, and the measurement of aqueous events including rain, hail, snow and fog with their associated temperature gradients provided later expeditions with helpful information that Parry had lacked.

The absence of thunder or lightning on or above the Arctic Circle surprised sailors and it was thought as a result that weather conditions nearer to the Pole were generally more settled compared with places below the Arctic circle. The levels of rainfall though could be extreme. The Aurora Borealis, or northern lights, were closely observed and its appearance correlated to any noticeable change to the magnetic variations on the compass. Its position in the sky and duration was thought to be associated with various factors including the orbit of the moon, temperature and easier weather conditions, but none of them could be explained or proved.

Parry, after many observations, was able to prove that the Aurora had no discernible effect on any of their different compasses nor the strength recorded with the dipping needle.

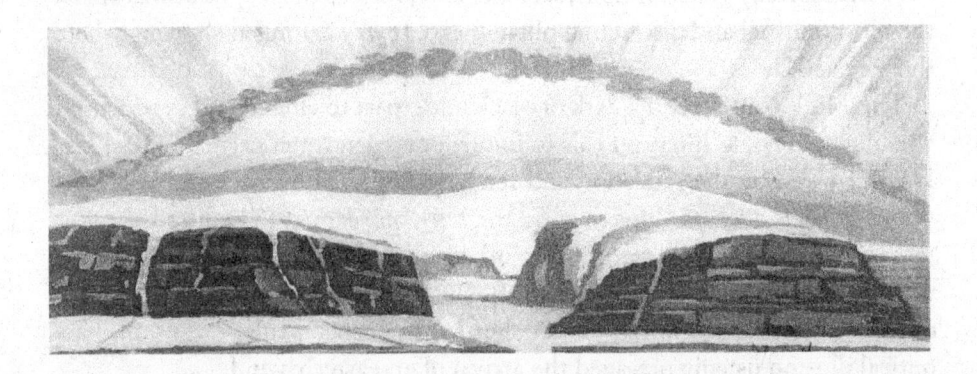

Notes

Chapter 1
1. A. Parry. *Parry of the Arctic. The life story of Admiral Sir William Edward Parry. 1790–1855.* Chatto and Windus. London. 1963, p.12.
2. C. Markham, Sir, KCB FRS *Lands of Silence* Cambridge University Press, Cambridge, 1921.
3. S. Lee (ed.) *Hobhouse. Benjamin.* Dictionary of National Biography. Vol.27. Smith, Elder and Co. London. Benjamin and Amelia had fourteen children. By his first marriage he had five, making nineteen children in total.
4. A full account of Caleb Parry's life and work is given in S. Glaser. *The Spirit of Enquiry. Caleb Hillier Parry MD FRS* A. Sutton 1995 ISBN.0750909986.
5. Caleb kept an extensive medical library. Some of his collection are held by the Bristol Medical School Library. Under *Parry Collection.*
6. J. Austen *Emma.* (Edited by J. Blythe) Penguin Classics London, 1966 note p.470.
7. See *Lives of British Physicians.* John Murray, London, p.830 and Jones B. Caleb Hillier Parry. West of England Medical Society. 1991. 106, p.101.
8. Madame D'Arblay. *Diaries and Letters 1842–1846* London, 1853.
9. E. Parry. Reverend. *Memoirs of Sir William Edward Parry KT FRS Late Lieutenant Governor of Greenwich Hospital.* London, 1858, Longman, Brown, Green, Longmans and Roberts, p.2.
10. SPRI MS.438/26/149-309. Letter to Charles Parry and (Fred) May 1799.
11. A. Parry. *Parry of the Arctic. The Life story of Sir William Edward Parry 1790–1855* Chatto and Windus, London, 1963, p.14.
12. E. Parry Reverend *Memoirs of Sir William Edward Parry KT FRS Late Lieutenant Governor of Greenwich Hospital.* London, 1858, Longman, Brown, Green, Longmans and Roberts, p.3. His son recounts an anecdote about his father atop a large Globe in a friend's library. When asked what he was doing replied, 'how I should like to go around it'.

Chapter 2
1. *Si vis pacem para bellum.* (Latin) Meaning If you want peace prepare for war.
2. R. Blake *Religion in the British Navy.1815–1859.* Chapter 2. Piety and Professionalism. One of his nicknames was 'Billy Blue' for this reason.
3. R. Morriss. (ed). *The Channel Fleet and the Blockade of Brest 1793–1802.* Ashgate Press, Navy Records Society. Aldershot, 2001 p.36
4. The Navy List was a process by which the Admiralty promoted senior officers depending on length of service and one position had to be vacated by death or retirement before a replacement was appointed. Some exceptions were allowed.
5. N.A.M. Roger. *The Command of the Ocean. A Naval History of Britain.1749–1815.* Allen Lane, London, 2004, p.530. 'Cornwallis was of very reserved habits and manners; of few signals and fewer words.'
6. SPRI M/S 438/26/149 1 July 1809.

7. S. Glaser. *The spirit of Enquiry. Caleb Hillier Parry. MD, FRS* Alan Sutton Publishing, Stroud, 1995, p.97.
8. Gantheaume had 21 sail of the line and Cornwallis 16. The French never let the protection of the shore batteries.
9. D. Cecil. *A Portrait of Jane Austen.* Penguin Books. Middlesex. U.K. 1980 p.101.
10. A Parry. *Parry of the Arctic. The life story Admiral Sir Edward Parry. 1790–1855.* Chatto and Windus, London, 1963, p.17
11. E. Parry (Rev) MA, *Memoirs of Rear Admiral Sir W. Edward Parry Kt. FRS* Longman, Brown, Green, Longmans, & Roberts, London, 1858, p.35. There is no name given to this brother officer.
12. SPRI M/S 458/26/11, Letter to Caleb and Sarah Parry, 29 May 1815.
13. SPRIM/S 438/26/11, Letter to parents 12 April 1815.
14. SPRI M/S 438/26/13, Letter to parents 3 May 1815.
15. SPRI MS. 438/26/10. Letter to parents from HMS *Vanguard* 13 August 1809. 'I have just been going on with Cowpers Poems. I never was so much delighted with anything in my life.' William Cooper (1731–1800) was a pre-Romantic poet and Parry eventually, spent two years living in his old house in Norfolk.
16. SPRI M/S 438/26/16. Letter to parents 4 July 1816.
17. SPRI M/S 438/26/16. Letter to parents 4 July 1816.
18. SPRI M/S 46/ Letter to parents. River St Lawrence. 4 July 1816.
19. SPRI M/S 438/26/16 Letter to Caleb Parry. 7 August 1816.

Chapter 3
1. A. Briggs. *The Age of Improvement 1783–1867.* Longmans, London, 1979, p.207. 'In 1816 and early 1817 there was gloom in the countryside and distress in almost all industrial areas'.
2. S. Glaser. *The Spirit of Enquiry. Caleb Hillier Parry MD FRS.* Alan Sutton Publishing, Stroud, 1995, p.84.
3. R Knight. *The Pursuit of Victory. The Life and Achievements of Horatio Nelson* Allen Lane. London, 2005, p 15
4. K. Austin. *The Voyage of the Investigator 1810–1803 Commander Matthew Flinders RN* Scolar Press, Aldershot, 1989. The Flinders expedition to Australia was one major survey only made possible because of the temporary truce between England and France. John Franklin was a junior officer on this voyage.
5. W. Scoresby Jnr. *An Account of the Arctic Regions; With a history and description of the Northern Whale Fishery. Volume 1.* Archibald Constable and Co Edinburgh, 1820, A comprehensive list from 861AD up to 1820 the time of W.E. Parry's first voyage Appendix III a.
6. G Williams, *Arctic Labyrinth. The Quest for the Northwest Passage* Penguin Books, 2010. Part 3, pp.169–266
7. J. Thompson, *The Seasons: Winter* This section for the first time represented beautiful aspects of winter and of the many beneficial effects that cold European climates had on the character of their populations and their way of living.
8. SPRI Franklin Ms. 248/394.
9. SPRI. Ms 438/26/18 Letter to parents from London 23 December 1817.
10. J.H. Brazell, London, Weather. HMSO London 1968. Appendix 1 p.153.
11. SPRI Ms 438/26/18. Letter to parents from London. 23 December 1817.

12. P. Berton. *The Arctic Grail. The Quest for the North West Passage. and the North Pole. 1818–1909.* Viking Penguin. NewYork. 1988. Sir George Hope. Lord of the Admiralty.. The patron and supporter of Captain John Ross.
13. J. Ross. *A Voyage of Discovery Made under the order of the Admiralty in his Majesty's Ships Isabella and Alexander for the purpose of Exploring Baffin Bay and enquiring into the probability of the North West Passage.* Longman, Hurst,Rees, Orme, and Brown, London, 1819. The number included 12 different Compasses, 7 Chronometers, Sextants, theodolites and Trengroves apparatus for saving lives!
14. SPRI. Ms 438/26/18. Letter to parents from London. 23 December 1817.
15. SPRI. MS.438/26/43. "My invaluable friend Maxwell. It is impossible for me to forget how much I owe him". 5 May 1819.
16. SPRI. 438/26/25. Maxwell writes to William about 'the great discussion that took place in the Admiralty over who should be appointed to lead the expedition' and a subsequent letter saying that he was 'made so happy by my appointment".
17. SPRI M/S 438/26/329. Letter to sisters 27 October 1832.
18. G. Williams, *Arctic Labyrinth. The Quest for the North West Passage* Penguin Books, London, 2010, p.172. Barrow was partly convinced by speculative geographers such as Buache, but more by Daines Barringtons FRS republished work of 1775, '*The Possibility of approaching the North Pole discussed*'.
19. The Eidophusician and other purpose built venues'. See Potter R. A. *Arctic Spectacles*. Washington University Press, 2007.
20. F.Spufford. *I May Be Some Time.* Faber and Faber, London, 1996, pp.58–61.
21. J. Conrad, *Geography and Explorers* Last Essays, London, 1926, p.14–16.

Chapter 4
1. Letters to Parry Family. 29 and 30 December 1817 MS 458/26/18/19. SPRI
2. ADM 55/4 NRO W.E. Parry. *Remarks on Magnetism and Meteorology 1818–1819.*
3. J. Ross. Captain. *A Voyage of Discovery Made under the order of the Admiralty in His Majesty's Ships Isabella and Alexander for the purpose of exploring Baffins Bay and enquiring into the probability of a North West Passage.* Longman, Hurst, Rees, Orme, and Brown, London, 1819. 'Few Voyages of this nature have excited more general interest at their outset than the present. Innumerable articles have appeared in the public journals which are in the hands of all classes of readers.' Intro p.3.
4. E. Parry. Rev. *Memoirs of Sir W. Edward Parry Kt. FRS* London 1857. Longman, Brown, Green, Longmans, and Roberts pp.69–70.
5. W. Barr & G. Williams. eds. *Voyages to Hudson's Bay in Search of a North West Passage Vol. 2* 1994
6. Ross. Captain. *A Voyage of Discovery Made under the orders of the Admiralty in his Majesty's ships Isabella and Alexander for the purpose of exploring Baffin Bay and enquiry into the probability of a North West Passaage.* Longman, Hurst, Rees, Orme, and Brown, London, 1819. 'Large collection of books taken including Cook but not Scoresby. Thirty Bibles and sixty testaments between both ships.' Intro p.xxi
7. SPRI. GB 15.MS 438/26/22. Dated 25 July 1819. Davis Strait.
8. Jonkers. A.R.T. *Earths Magnetism in the Age of Sail.* John Hopkins University Press, 2003. In the seventeenth and eighteenth centuries, the geomagnetic field was even less well understood than at present. Avigil had to be kept through frequent observation stored in the ships navigational log books and allowance made for needle deflection from true north in steering, dead reckoning and charting. Back at home, hydrographers compiled

tables of measurements and updated sailing directions to reflect the field's alteration. pp.3, 4.
9. J. Ross. K.S. *A voyage of Discovery Made under the Order of the Admiralty in His Majesty's Ships Isabella and Alexander for the purpose of exploring Baffins Bay and enquiring into the probability of a North West Passage.* Longman, Hurst, Rees, Orme and Brown .London 1819 p.50.
10. E. Hooper. RN. (Purser). *Voyage of the Isabella and Alexander 1818. Private Journal.* Entry 26 May. 'The first icebergs compared with angels wings, extended over the shoulders of some human forms indicating an ethereal Being. Exquisite blueish green' p.35.
11. W.E. Parry. Lieutenant Commander. *Letters of proceedings and weekly reports of Observations while commanding HMS Alexander addressed to Captain John Ross Commanding HMS Isabella during the Arctic Voyage of 1818.* National Archives of Canada A822. *ref SPRI*
12. J.Ross. K.S *A Voyage of Discovery made under the Orders of the Admiralty in His Majesty's Ships, Isabella and Alexander for the purpose of exploring Baffins Bay and enquiring into the possibility of a North West Passage. Vol. 2. Instructions.*
13. J. Ross. KS *A Voyage of Discovery Made under the orders of the Admiralty in His Majesty's Ships Isabella and Alexander for the purpose of exploring Baffin Bay and enquiring into the possibility of a Northwest Passage.* Longman, Hurst, Rees and Browne London.1819 Vol. 2 p.75 Insructions to Captain J.Ross. 'You must observe effects of refraction by observing an object either terrestrial or celestial over a field of ice as compared with objects observed over a surface of water'..
14. E.Hooper. *Voyage of the Isabella and Alexander. 1818.* Private Journal Vol. 1. R. G. S p.213 xv E. Hooper. *Voyage of the Isabella and Alexander 1818.* Private Journal. Vol. 1. R G S p.223.
15. J.Ross. K.S *A Voyage of Discovery Made under the orders of the Admiralty in His Majesty's Ships Isabella and Alexander for the purpose of Exploring Baffins Bay and enquiring into the possibility of a North West Passage'.* Longman, Hurst, Rees, Orme and Browne, London, 81p Vol. 2 p.65 Instructions. 'you are to cause views of bays harbours, headlands etc to be carefully taken.'
16. John Sacheuses sketch was adapted as a coloured engraving giving Ross the opportunity to embellish the rubbing noses story.

Chapter 5

1. SPRI MS. 4398/26/23 Letter to parents from Northumberland avenue Coffee House. 20 January 1819.
2. SPRI MS. 4398/26/24 Letter to parents. 21 January 1819.
3. SPRI MS. 438/26/25. Letter to parents. 22 January 1819.
4. SPRI MS. 438/86/27. Letter to parents 30 January 1819.
5. SPRI MS. 438/26/34. Letter to parents 25 February 1819.
6. SPRI MS. 438/26/35. Letter to parents. 27 February 1819. 'I fear I have half offended some who are not considerate enough to see the true cause but I cannot help it.'
7. SPRI MS 438/26/38. Letter to parents. Deptford. 12 April 1819
8. SPRIMS 438/26/37. Letter to parents. Deptford. 5 April 1819.
9. RGS. RGSU 21382SSCc/73. Hooper Collection. W.E. Hooper. Private Journal. Volume 2. 29 June 1819.
10. SPRI MS438/26/22. Letter to parents. Davis Strait. July 25 1819.

Chapter 6

1. W. E. Parry. RNFRS *Three Voyages for the Discovery of a North West Passage from the Atlantic to the Pacific.* John Murray, London, 1821, p.47.
2. W.E. Parry. RNFRS Journal of a *Voyages for the Discovery of a North West Passage from the Atlantic to the Pacific* John Murray, London, 1821, p.36
3. W.E.P. Hooper. *Private Journal of E. Hooper. Vol. 2.* RGS Library, p.253. Hooper mentions that Parry ordered extra rations for those men working with the saws.
4. W.E.P. Hooper. *Private Journal of E Hooper. Vol. 2.* RGS Library Hooper Collection rgsu 21382 SSC/73. Hooper lists eight plays performed one of which Parry wrote himself called 'The North West Passage' or 'Voyage finished', a clear attempt to keep up motivation for the task ahead!
5. Both Hoppner and Beechey were artistically capable of representing different weather phenomena associated with polar conditions; optical effects such as inverted icebergs, a water sky and the 'blink'. The Aurora's movements and colours however appears to have confounded them.
6. W.E.P. Hooper. *Private Journal. Vol 1.* Library of the RGS. Hooper Collection. RGSU. 21382 SSC/73 'The extraordinary effects of frost on a man's faculties are extremely curious, resembling drunkenness in every respect and it is quite impossible to distinguish by which of these causes he is affected. He looks wild speaks thick and incoherently so he cannot give a rational answer to any question put to him'. Section 2, p.11
7. W.E. Parry RN FRS. *Journal of a Voyage for the Discovery of a North West Passage from the Atlantic to the Pacific* John Murray, London, 821, p.125.
8. He wrote and had published a pamphlet on this subject Sir W.E. Parry.. *A lecture on the Character, Condition and Responsibilities of British Seamen* Harrison and Son, London, 1854.
9. C. S Hastings *Monthly Weather Review* Washington, 920 p.22.
10. W.E. Parry. RNFRS *Journal of a Voyage for the discovery of a North West Passage from the Atlantic to the Pacific* John Murray, London, 1821, p.103.
11. M/s 435/36/49 Letter to Parent's. Davis Strait. 5 September 1820.

Chapter 7

1. Anthony. Earl of Shaftesbury (Ed. J.M. Robertson) *The Moralists* in; *Characteristics of Men, Manners, Opinions Times* NewYork, 1964, p.119.
2. Arguably the most influential work was James Thomson poem in *The Seasons – Winter 1719* and republished throughout the eighteenth century. This shifted perceptions of winter climates and the cold regions of the world towards an idea that they produced stronger, braver and more resourceful human beings than warmer countries in the south.
3. Chauncey C. Loomis. *The Arctic Sublime.* In Nature and the Victorian Imagination. University of California Press, Berkeley. 1977, p.99.
4. A. Fisher. *A journal of a Voyage of Discovery to the Arctic Regions in the years 1819 and 1820* Longman, London, 1821.
5. SPRI GB 15. Sir Edward Sabine 1788–1883 led a very distinguished life. He was awarded the Royal Society Copley medal for his pendulum work during Parry's first Expedition 1819–1820. He made the first systematic magnetic survey of the British Isles and became President of the Royal Society in 1861. He was promoted to General in 1870.
6. M/S 435/26/54 Letter to Parents London. 6 December 1820.
7. C. Markham.Bt. KCB FRS *Lands of Silence.* CUP, London, 1921, p.76.

8. G. Back Bt. *Narrative of an Expedition in HMS Terror undertaken with a View to Geographical Discoveries on the Arctic Shores. 1836–1837* John Murray, London, 1838, p.276.
9. MS 435/26/54 Letter to Parents. London, 6 December 1820
10. A. Lambert *Franklin; Tragic Hero of Polar Navigation*. Faber and Faber London, 2009, p.36. An Imperial Russian Decree to this effect was passed on 28 September.1821.
11. M/S 435/26/54 Letter to Parents. 6 December 1820..
12. SPRI M/S 435/26/54. 6 December 1820. Letter to parents. He and Sabine were elected Honorary Members of the Royal Society at this dinner.
13. F. Fleming. *Barrow's Boys*. Granta Books. 1998 p.98.
14. *Bath Chronicle and Weekly Gazette* Saturday, 29 March 1821.
15. *Bath Chronicle and Weekly Gazette* 14 March 1822, Decease of Caleb Parry MDFRS
16. SPRI M/S 438/26/52. Letter to Sarah Parry. 'A Mr Parkins was the printer and had prepared the engravings of his charts on 'steel'. 'This innovation may enable copies to be made with facility, beauty and quickness. 100,000 copies are possible!'
17. Captain W.E. Parry. FRS *Journal of a Discovery of a North West Passage from the Atlantic to the Pacific Oceans Performed in His Majesty's Ships Hecla and Griper 1819–1820* John Murray, London, 1821 Intro xvi
18. Sir John Barrow. Bart. FRS *Voyages of Discovery and Research within the Arctic Regions from the Year 1818 to the present time* John Murray, London, 1846
19. Captain W.E. Parry. FRS *Journal of a Discovery of a North West Passage from the Atlantic to the Pacific Oceans. Performed in His Majesty's Ships Hecla and Griper under the orders of William Edward Parry RN FRS 1819–1820* John Murray, London, 1824, Intro pp. v–vi.
20. Captain W.E. Parry FRS, *Journal of a discovery of a North West Passage from the Atlantic and Pacific Oceans in his Majesty's Ships Hecla and Griper*. Official Instructions 27April 1821 pp.xxi-xxx.
21. Captain W.E. Parry FRS, *Journal of a discovery of a North West Passage from the Atlantic to the Pacific Oceans in His Majesty's Ships Fury and Hecla 1821–1822–1823* John Murray, London, 1824, Official Instructions, p.,xxi – xxx.
22. M/S, 430/26/496 SPRI Letter to Parents. Undated.

Chapter 8
1. G.F. Lyon. *Private Journal of Captain G.F. Lyon of HMS Hecla. During the recent voyage of discovery under Captain Parry* John Murray, London, 824, p.30.
2. G.F. Lyon. *Private Journal of Captain G F. Lyon of HMS Hecla during the recent voyage of Discovery under Captain Parry* John Murray, London, 1824, p.234
3. G.F. Lyon. *The Private Journal of Captain G.F. Lyon* London, 1824 'The kind and cheerful manner in which they exerted themselves to contribute to our comfort and amusement on all occasions when at the huts, will long be cherished in our memories when all their faults shall be forgotten and we have ceased to think of them in any other light than they really are – a kind hearted, peaceable and estimable people.' p.527.
4. G.F. Lyon. *The Private Journal of Captain G F Lyon* London, 1824.
5. W.E. Parry. *Journal of a Discovery of a North West Passage from the Atlantic to the Pacific. Performed in His Majesty's Ships Fury and Hecla under the orders of Captain William Edward Parry. RNFRS* John Murray, London, 1824, p.118.
6. Captain W.E. Parry, *Journal of a Discovery of a North West Passage from the Atlantic to the Pacific. Performed in His Majesty's Ships Fury and Hecla under the orders of Captain William Edward Parry. RNFRS* John Murray, London, 1824, p.123.

7. The Admiralty Instructions had laid emphasis on the need for a study of the effect of the Aurora Borealis on Electromagnetic and Magnetic needles.
8. G.F. Lyon. *The Private Journal of Captain G.F. Lyon* John Murray, London, 1824. Lyon reported that the Inuit spoke about death with a 'degree of solemn awe and terror that seemed to imply a dread of the event. Iligliuk said they went into the earth after death but Amaneelia that they ascended to the skies.' Among many Inuit the Aurora when it appeared, was a sign of the spirits of their 'elders'.
9. G.F. Lyon. *The Private Journal of Captain G F. Lyon* John Murray, London, 1824, p.356.
10. W. Hooper. *Private Journal 3 Volumes*. RGS.
11. SPRI M/S 1021.Douglas Clavering letters. There is no real proof of this other than a report by J.C. Ross on board *Hecla*, who reported it to Douglas Clavering an old naval friend.
12. G.F. Lyon. *The Private Journal of Captain G.F. Lyon* John Murray, London, 1824 p.127
13. William Hooper. *Private Journal* 3 Volumes RGS. pp.176–178.
14. K. McGoogan. *Fatal Passage. The untold story of Scotsman* John *Rae the Arctic Explorer who discovered the fate of Franklin*. Bantam Books London 2002 pp.16,17,18 have vivid account from Rae of the condition and the available remedies.
15. SPRI M/S 1023 Letter to Maxwell. 24 October 1823.
16. G.F. Lyon. *The private journal of Captain G.F Lyon* John Murray, London, 1824 p.439.
17. SPRI MS. 438/86. Letter to Charles Parry. January 1824. Parry was respectful and praising of George Lyon and James Clark Ross. The other officers may or may not have been guilty of 'quarrels, insubordination and misbehaviour'. as suggested by Clavering.
18. E. Parry. *Memoirs of Rear Admiral Sir W. Edward Parry. Kt. FRS* Longman, Brown, Green, Longmans & Roberts. London. 1858 p.152–153
19. Captain W.E. Parry. F R.S. *Journal of a Voyage for the Discovery of a North West Passage from the Atlantic to the Pacific performed in the years 1821–22–23 in HM Ships Fury and Hecla under the orders of Captain William Edward Parry* John Murray, London, 1824, p.491.
20. J. Barrow. Kt FRS, *Voyages of Discovery and Research within the Arctic Regions from the year 1818 to the present time. Several Naval Officers employed by Sea and Land in search of a North West Passage from the Atlantic to the Pacific, with two attempts to reach the North Pole* John Murray, London, 1846 p.211.
21. Mellon Institute G650 1821 p.25 Captain G. F. Lyon.1795–1832.Author. Illustrator. *The Private Journal of Captain George Francis Lyon of HMS Hecla during the recent voyage of Discovery under Captain Parry*. 468 pages. Seven leaves of plates.

Chapter 9

1. G. Williams. *Arctic Labyrinth. The Quest for the North West Passage*. Penguin Books, London, 2010. Williams quotes C. Stuart Houston who compiled the various manuscript journals of Franklin's companions Hood, Back and Richardson. 1974. 'The forgotten martyrs of the expedition. They carried two packs each weighing 75 to 90 lbs. on their backs and were often with nothing or little to eat.' p.207.
2. A. Brandt. *The Man Who Ate His Boots. Sir John Franklin and the tragic History of the North West Passage*. Jonathan Cape, London, 2011. Barrow is quoted by the author as saying 'a beautiful example of the triumph of mental and moral energy over mere brute strength, in the simple fact that, out of twenty individuals in the group fifteen Voyageurs sank down and died and only one English seaman unaccustomed as they were to the severity of the climate, succumbed, and he by the murderous hand of an assassin.' p.141.

3. A. Lambert. *Franklin. Tragic Hero of Polar Navigation.* Faber and Faber, London, 2009 p.41.
4. F. Spufford. *I May Be Some Time.* Faber and Faber, London, 1996, p.52 Franklin was part of a Lincolnshire circle that included banks and several other large and wealthy landowners. He was related to Tennyson, his nephew in law, and became engaged to Eleanor Porden a poet and member of the Institut de France.
5. SPRI MS 438/26/157. Letter to Charles Parry. 20 November 1823. 'I have seen much of my worthy friend Franklin. He like myself has got another trip in his head. To ask the Hudson's Bay Company to back him on a trip to the McKenzie River as a "commercial expedition" as the Russians bid fair to exclude us from the most valuable (i.e. NW) parts of N. America if we do not by prior visits and discovery establish a claim.'
6. A. Parry. *Parry of the Arctic. The Life story of Admiral Sir Edward Parry. 1790–1855.* Chatto and Windus, London, 1963, p.96
7. W.E. Parry. Letter to Gertrude Parry. Parry Family Archive.
8. W.E. Parry. *Journal of a Voyage of Discovery of a North West Passage from the Atlantic to the Pacific performed in the years 1821–1822–1823 in HM ships Hecla and Fury under the orders of Captain W.E. Parry* John Murray, London, Appendices. 1824a. The new plants were *Eutrema Edwardsii, the Parrya, and Pleuropogon Sabinii.* Robert Brown had been Sir Joseph Banks' librarian for many years.
9. W.E. Parry. *Journal of a Voyage of Discovery of a North West Passage from the Atlantic to the Pacific performed in the years 1821–1822–1823 in HM Ships Hecla and Fury under the orders of Captain W.E. Parry* John Murray, London, 1824 p.xviii.
10. SPRI MS 438/12/10.
11. Franklin was a strict Sabbatarian according to his wife, Jane Porden. So strict they had several disagreements about Sunday activities and other principles.
12. SPRI M/S 438/26/154. Letter to mother and sisters. October 18 1823. 'God help you all dearest ones, which is all that I can add at present'. His fathers death was 'a painful dispensation of providence'.
13. A. Parry. *Parry of the Arctic. The life story of Sir Edward William Parry 1780–1855.* Chatto and Windus, London, 1968 pp.83 -84 Caroline was very happy and content with Joseph Martineau and they settled at Stamford Hill just north of London. Parry spent many nights with them between his voyages.
14. MS *Letter to brother Charles from Admiralty. 5/1/1824.* SPRI
15. E. Parry. *Memoirs of Rear Admiral Sir Edward Parry Kt.* Longman, Brown, Green, Longmans and Roberts, London, 1858 p.181.
16. W. Hooper. *Journal of 1824–1825.* MS. RGS London, pp.125–7.
17. L.N. Rosen, SD Targum, M. Terman, M. J. Bryant, H. Hoffman, S. F. Kasper, J. R. Hamovit, J. P. Docherty, B. Welch, N.E. Rosenthal. *Prevalence of Seasonal Affective Disorder at Four Latitudes.* Psychiatry Research 31:131.-144. Elsevier Publishing Ireland Ltd.1989. 'Problems appear to vary and increase in direct proportion to Latitude but bore no relationship to Longitude.' p.131.
18. M/S 438/26/154. *Letter to Mother and Sisters. from Admiralty.* 18/10/1823. SPRI.
19. SPRI 231/21/54 M. Crozier. *Private copybook of Orders. HMS Hecla.* 19 August 1825.
20. WE Parry. *Journal of a Voyage of Discovery of a North West Passage from the Atlantic to the Pacific performed in the years 1823–24 in HM Ships Hecla and Fury under the orders of Captain W.E. Parry* John Murray, London, 1825 p.164.
21. For an account of this remarkable escape the reader is recommended two very different versions., J. Ross and J. C. Ross. *Narrative of a Second Voyage in search of the North West*

Passage John Murray, London, 1835. and R. Huish. *The Last Voyage of Capt. Sir John Ross*, London, 1835.
22. G. Williams. *Arctic Labyrinth. The Quest for the North West Passage*. Penguin Books, London, 2009. Amundsen completed the passage in a shallow draught yacht the *Gjoa* in August 1906. She started in 1803 pp.365–9.

Chapter 10
1. SPRI. M/S 438/26/248. Letter to Sarah Parry. 16 Octtober 1825.
2. SPRI. M/S 438/26/168. Letter to Charles Parry. 22 October 1825.
3. SPRI.M/S 438/26/170. Letter to Charles Parry. 24 October 1825.
4. SPRI MS 438/26/171. Letter to Charles Parry. 25 October 1825.
5. SPRI. M/S 438/26/170. Letter to Charles Parry 27 October 1825.
6. J. Adeane (ed) *Catherine Edwards Parry; a record of her life* Longmans Green, London, p.37
7. SPRI M/S 438/26/163 Letter to Charles. 'The Fry's are really most superior people'.
8. SPRI M/S 438/26/163.Letter to Charles. The Fry's and their circle comprised many Quakers and her husband was a banker. On neither count did they appeal to Lady Stanley.
9. J. Adeane. *The Girlhood of Maria Josepha Holroyd* and, *The Early Married Life of Maria Josepha Lady Stanley* Longmans Green, London, 1896 and 1899.
10. A. Parry. *Parry of the Arctic*. Chatto & Windus, London, 1963, p.99.
11. A. Parry. I Parry. *Unpublished Journal*. Parry Family archive.
12. SPRI MS 438/26/256. Letter to Mrs Sarah Parry. 21 March 1826. 'A tolerably pleasant party at the Duke's yesterday. Lords Morpeth, Prudhoe, Seymour and Bexley and Prince Leopold. I was rather out of my element.'
13. SPRI M/S 578/1. Letter to Mrs Sarah Parry 23 October 1826
14. A. Parry, *Parry of the Arctic*. Chatto and Windus, London, 1963, p.102. Extract from Isabella Parry's Journal.
15. SPRI M/S 438/26/260. Letter to Charles Parry. 7 June 1826.
16. Barrow, Sir, FRS *Auto-biographical Memoir of Sir John Barrow, Bart, late of the Admiralty including reflections, observations and reminiscences at home and abroad from early life to advanced age* John Murray, London, 1847, p.324. 'Actual measurement of the meridian at the Pole and the Equator would repay any effort that might be made to effect it.' and p.321, 'An observer at the Pole being on no one meridian but at a point where all meridians meet apparent time has no longer either existence or meaning.'
17. P. Berton. *The Arctic Grail. The quest for the North West Passage and the North Pole. 1818–1909*. Viking Penguin Group, New York, 1988, p.97.
18. SPRI M/S 438/26/262. Letter to Sarah Parry 19 July 1826.
19. SPRI M/S 438/26/160. Letter to Charles Parry. 'I am living at the rate of £2,000 per annum but I believe it is worth the money.' This was before his marriage to Isabella. Anne Parry suggests his income was approximately £850 per annum including the £500 he received for the Hydrography sinecure but excluding any publishing contracts he received.
20. The best advice came from William Scoresby Jnr. who believed that light sledges drawn by dogs were the best means of Arctic travel and that the route should be from the west not the east. He had been as far north as 82deg 30' which was an unofficial record and further than any other explorer.
21. E Parry Rev. *Memoirs of Rear Admiral Sir Edward W Parry Kt*. Longman, Brown, Green, Longmans, & Roberts, London, 1858 p.191. 'The form of commissioning is merely hoisting the pendant and when a ship is paid off the same is hauled down. She (Isabella)

hoisted it and I saw her hoist it. Everybody who was by was quite delighted. This is just what seamen delight in'.
22. A. Parry. *Parry of the Arctic.* Chatto & Windus, London, 1963, p. 104.
23. I. Parry, *Unpublishd Manuscript Journal. August 1826 – March 1827.* SPRI MS.438/4/1. BJ.
24. F. Spufford. *I May be Some Time.* Faber & Faber, London, 1996, p.100.
25. I. Parry. *Unpublished Manuscript Journal. August 1826 – March 1827.* SPRI M.S.438/4/1. 'I may not accompany my own dear husband, but Fido may accompany his Master and comfort him and speak to him of his Mistress therefore with ENVY I send him at the request of my dear Man.'
26. SPRI M/S 438/26174. Letter to Charles Parry 16 March 1827
27. SPRI MS438/26/175. Letter to Charles from the Nore. 31 March 1827.
28. A. Parry. *Parry of the Arctic.* Chatto and Windus, London, 1963, p.106.

Chapter 11
1. W.E. Parry. *Narrative of an attempt to reach the North Pole in Boats fitted for the purpose and attached to HMS Hecla in the year MDCCCXXVII* (1827) *under the command of Captain W.E. Parry RN FRS* John Murray, London, 1828, pp.283 and 294.
2. W.E. Parry. *Narrative of an attempt to reach the North Pole in Boats fitted for the purpose and attached to HMS Hecla in the year MDCCCXXVII.* (1827) *under the command of Captain W.E. Parry RN FRS* John Murray, London, 1828.
3. W.E. Parry. *Narrative of an attempt to reach the North Pole in Boats fitted for the purpose and attached to HMS Hecla in the year MD CCCXXVII* (1827) *under the command of Captain W.E. Parry RN FRS* John Murray, London, 1828.
4. W.E. Parry. *Narrative of an attempt to reach the North Pole in Boats fitted for the purpose and attached to HMS Hecla in the year MDCCCXXXVII (1827) under the command of Captain W.E. Parry RN FRS* John Murray, London, 1828.
5. C.R. Markham. Bt. KCB FRS *Lands of Silence,* Cambridge University Press, London, 1921.
6. W.E. Parry, Narrative of an attempt to reach the North Pole in Boats fitted for the purpose and attached to HMS *Hecla.* in the year MDCCCXXXVII. Under the Command of Captain W.E. Parry. RN FRS, John Murray, London, 1828 p.200.
7. E. Parry M.A. *Memoirs of Rear Admiral Sir W. Edward Parry, KT FRS,* Longman, Brown, Green, Longmans, & Roberts, London, 1858., p.201.
8. SPRI. M/S 438/26/273. Letter to Isabella Parry. 27 September.1827.
9. SPRI. M/S 438/26/274.Letter to J.C. Ross 30 September 1827.
10. SPRI 438/26/179. Letter to Charles. 25 October 1827.
11. SPRI M/S 438/26/178.
12. SPRI M/S 438/26/177. Letter to Charles Parry. 8 October 1827.
13. SPRI M/S 438/26/279. Letter to Sarah Parry. 4 November 1827.
14. W.E. Parry. *Narrative of an attempt to reach the North Pole in Boats fitted for the purpose and attached to HMS Hecla in the year* MDCCCXXVII *under the command of Captain W.E. Parry RN FRS,* John Murray, London, 828.

Chapter 12
1. A. Parry, *Parry of the Arctic. The Life story of Admiral Sir Edward William Parry 1790–1855,* p.128.
2. E. Rigby, *Memoirs of Lady Eastlake* John Murray, London, 1849. Rigby was a cousin of Parry. She described Croker as 'singularly entertaining and disagreeable.'
3. L.S. Dawson. Cmdr. RN *Memoirs of Hydrography, 1750–1885.*

4. NMMM SS/78/038 T. Croker. *Letter to Editor of London Gazette. 3/04/1828*. His letter includes mention of Captain Owen's survey of Africa and the use of the *Hecla*. Parry was strongly against sending his old ship into tropical waters which, he predicted, would be the finish of her. He tried to intervene and quarrelled with Croker. *Hecla* never came back.
5. SPRI M/S 438/26/279. Letter to Isabella. 4 November 1827
6. A.H. Markham. *The Life of Sir* John *Franklin, London,* 1891
7. A. Parry. *Parry of the Arctic. The Life story of Admiral Sir Edward William Parry 1790–1855*, p.133. It seems clear that Isabella had written to her mother confessing that she loved her little son too well and she rejoiced that her children had been removed from the wicked world to join the angels. Lady Maria interpreted this state of Christian resignation on Parry's influence but Isabella's religious nature suggests it was entirely her own conviction.
8. SPRI.
9. F. Fleming. *Barrow's Boys*. Granta Books, London, 1998, pp.276–277. Fleming aptly describes the rest of Australia outside New South Wales as 'a geographical attic'!
10. SPRI. M/S 434/26/279. Letter to Sarah Parry. Bishopstoke, 4 November 1827. He refers in delight to an advance of 300 guineas for his book on the expedition to the Pole. 'more than I expected'. He was extremely short of money at this time.
11. The Oxford University DCL prize poem was ten verses long.
 'But Fairer England greets
 the wanderer now,
 Unfaded laurels shade her Parry's brow, and on the proud memorial of her fame,
 lives, linked with deathless glory, Franklin's name.
12. *Berkshire Chronicle* 5 May 1847. Sir John Eardley-Wilmot married Elizabeth Chester in 1818. He was an MP in Warwickshire before becoming Lieutenant Governor of Tasmania in 1843
13. SPRI. Letter to Charles Parry 3/6/1829 Ms 438/26/186.

Chapter 13

1. The Royal Australian Historical Society. *The First Decade of the Australian Agricultural Company Mit*chell Library, NSW The Royal Charter of 1/1/1824 laid out clear terms on which the AAC held their Grant. This included the employment after twenty years of 1,400 convicts, and that £10,000 had to be spent in the formation of roads, clearing, fencing and draining the selected lands.
2. *The Sydney Gazette* 23 October 1827.
3. Parry kept a daily journal throughout his stay in New South Wales. The original is in the Mitchell Library, Sydney. A transcript by one of his great-grandchildren is in the family archive.
4. Sir E.W. Parry, *Australian Journal 1829–1834* Mitchell Library, Sydney. Entry for 28 December 1829.
5. Sir E.W. Parry. *Australian Journal.1829–1834 Mit*chell Library, Sydney Entry for 5 January 1830. 'It is believed that many of the proprietors (shareholders) will forfeit their shares rather than pay! Nothing can more strongly prove the utter want of money in the Colony'.
6. Sir W.E. Parry. *Australian Journal.1829–1834* Mitchell Library, Sydney. Entry for 10 January 1830
7. Parry Family Collection. Letter to Sarah Parry. 19 January 1830.
8. Sir E.W. Parry. *Australian Journal. 1829–1834 Mit*chell Library, Sydney. 'The sheep were in flocks of 300 to 500, each watched by two shepherd and a hut keeper, all convicts.' p.150

9. Sir W.E. Parry *Australian Journal.1829–1834* Mitchell Library, NSW. 11 March 1830.
10. Parry Family Collection. Letter to Sarah Parry. 24 July 1830.
11. Sir E.W. Parry. *Australian Journal. 1829–1834* Mitchell Library, NSW. 'Mr Barton's Manner and tone are peevish, discontented, complaining and even to Me a stranger, More than half insolent.'
12. Sir E.W. Parry. *Australian Journal.1829–1834* Mitchell Library, NSW p.157.
13. Sir E.W. Parry. *Australian Journal.* 1829–1834 Mitchell Library, NSW p.42. Entry for 21 December 1830.

Chapter 14

1. A. Brandt. *The Man Who Ate His Boots. Sir* John *Franklin and the Tragic History of the North West Passage* Jonathan Cape, London, 2011, pp.234–235.
2. A. Parry, *Parry of the Arctic. The Life Story of Admiral Sir William Edward Parry 1790–1855* Chatto and Windus, London, p.162. The other key members of the Republican group were Messrs Keith, Francis Stephen, Hall and Haynes. These individuals arranged a disorderly Fete on Pinchgut Island to celebrate the Darling's departure.
3. A. Parry, *Parry of the Arctic. The Life Story of Admiral Sir William Edward Parry 1790–1855.* Chatto and Windus, London, 1963, p.163. Parry wrote later 'he entered freely on the affairs of the Colony and the Company and seems really desirous of obtaining every information respecting them.'
4. M/S 438/26/188. Letter to Charles Parry. 14 July 1832. SPRI
5. Captain W.E. Parry, *Australian Journal 1829–1834.* Parry Family Archive. Entry for August 25th 1831, p.65.
6. Captain W.E. Parry, *Australian Journal 1829–1834.* Parry Family Archive. Entry for 15 March 1832, p.84.
7. Captain W.E. Parry, *Australian Journal 1829–1834.* Parry Family Archive. Entry for 22 September 1832, p.98. His decision was contrary to his own domestic happiness since at Tahlee, near Carrington, he had recently planted a vineyard, a lawn, and many new rosebushes.
8. A. Parry. *Parry of the Arctic. The Life Story of Admiral Sir William Edward Parry 1790–1855.* Chatto and Windus, London, *1963, p.175.*
9. M/S438/26/189, Letter to Charles Parry.19 December 1832. SPRI
10. Captain W.E. Parry, *Australian Journal 1829–1834. Parry Family Archive p.130.* He followed this with one last visit to Booral.
11. E. Parry. Reverend, *Memoirs of Rear Admiral Sir Edward Parry Kt. FRS etc. 4th Edition* London, Longman, Brown, Green, Longmans, and Roberts, 1858, p.242.

Chapter 15

1. G. Clark.& M.E. Page. *Welfare Reform, 1834. Did the New Poor Law in England produce significant economic gains?* Cleometrica. 13. (2) ISSN 1863–2505 world cat .org/issn/1863 -2505, pp.221–44. 2019. This recent study concluded that 'the 1834 Welfare Reform had little or no impact on rural wages, labour mobility, or the fertility rate of the poor. The study concludes, 'this deliberately induced suffering gained little for the land and property owners who funded poor relief. Nor did it raise wages for the poor or free up migration to better opportunities in the cities. One of the first great triumphs of the new discipline of Political Economy, the reform of the Poor Laws, consequently had no effects on economic growth and economic performance in Industrial Revolution England.'

Notes   213

2. SPRI. MS 438/26/333.Letter to sisters.27. 111 1834. He said he had declined because of ability (none) and ill health. He also referred to a desire for a 'quiet backwater' where he could recover his strength and health.
3. SPRI. MS 438/26/192 Letter to Charles. 8 December 1834.
4. SPRI. MS 438/25/195. Letter to Charles. 31 January 1835.
5. A. Parry. *Parry of the Arctic. The Life Story of Admiral Sir Edward Parry 1790–1855.* Chatto and Windus, London, 1963, p.180.
6. Parry Family Archive. Letter to Charles 23 February 1836. 'You will not wonder that I should be sick of the sight of pen and ink, when I tell you that, for months past I have been writing about sixty letters a week by filling up the intervals between other duties; the latter involving the travelling of 1,600 miles a quarter, chiefly in a gig.'
7. A. Parry. *Parry of the Arctic. The Life Story of Admiral Sir Edward Parry. 1790–1855.* Chatto and Windus, London, 1963, p.185.
8. A. Briggs. *The Age of Improvement 1783–1867.* Longmans, London, 1996 pp.282–3.
9. F. Fleming. *Barrow's Boys.* Granta, London, 1989 p.249.
10. J. Ross. *Narrative of a Second Voyage in Search of a North West Passage during the years 1829, 1830, 1831, 1832, and 1833* Webster, London, 1835, p.292
11. G. Williams. *Arctic Labyrinth. The quest for the North West Passage.* Penguin Books, London, 2010, p.247. James Ross refuted his uncle's claims by writing to him, 'my claim of the exclusive honour of the discovery of the Magnetic Pole .... The merit of the discovery is to me *alone*. I wish to receive from you an explicit admission of that fact.'
12. SPRI. MS. 438/26/196. Letter to Charles. 23 May 1835.
13. SPRI. MS 438/26/196. Letter to Charles. 23 May 1835.

Chapter 16

1. A. Parry. *Parry of the Arctic. The Life Story of Admiral Sir Edward Parry 1790–1855* Chatto and Windus, London, 1963, p.197. 'The addition of between 20 and 30 steam vessels to those already belonging to the Navy by the transference of the Post Office packets to the Admiralty only calls for the immediate attention of the Board to a measure, the necessity of which has been growing more apparent for some time, viz. the providing of the more efficient control and superintendence of what may now be fairly called the Steam Department of the Naval Service.'
2. A. Parry. *Parry of the Arctic. The Life Story of Admiral Sir William Edward Parry 1790–1855.* Chatto & Windus, London, 1963, p.197.
3. SPRI. M/S 78/038. *Autobiographical Memoir of Robert Second Viscount Melville. 1828–1830*
4. F.C. Smith. Capt. OBE RN, *A Short History of Naval and Marine Engineering.* Cambridge University Press, Cambridge 1938. The Lightning was 126ft long, 296 tons and 'the first steam vessel in the British Navy to take part in Naval Warfare and the first for which a commission was granted'.
5. J.H. Briggs. Kt. *Naval Administrations. 1827–1892* Sampson, Low, Marston & Co, London, 1897, p.6.
6. E.C Smith Eng. Capt. OBE RN, *The Centenary of Naval Engineering* Lecture at the Institute of Marine Engineers, London, 30 March 1922 p.90.
7. J. H. Briggs Kt. *Naval Administrations 1827–1892* Sampson, Low, Marston & Co, London, 1897, p.15.
8. Sir John Barrow Bart. *An Autobiographical Memoir* John Murray, London, 1847, p.76
9. G. Davies. Marine Quarterly. Issue Summer 2021. *The Iron Ships era* pp.65–74. This is a brilliantly condensed article about the early days of iron shipbuilding and the steam engines that powered them.

10. E.C. Smith. Eng. Capt. OBE RN. *The Centenary of Naval Engineering* Lecture at the Institute of Marine Engineers, London, 30 March 1922, p. 91.
11. *The Sun*. Tuesday 19 June 1855. The announcement by the Provisional Committee for a 'suitable testimonial to Mr Francis Petit Smith for his eminent services in bringing into practice the system of screw propulsion'. It raised in excess of £4,000.
12. I Brunel. *Life of Isambard Kingdom Brunel, London*, 1870. By this time Brunel had already decided to convert his great iron ship SS *Great Britain* to screw propulsion. He only agreed to work with the Admiralty on the *Rattler* if he dealt with Parry directly and any Board members he alone suggested. This was to circumvent Symonds, although he managed nonetheless to delay the building of the ship for several years.

Chapter 17

1. A. Parry. *Parry of the Arctic. The Life story of Admiral Sir William Edward Parry 1790–1855* Chatto and Windus, London, 1963, p.196.
2. E. Parry, Rev. *Memoirs of Sir Edward Parry Kt FRS* Longman, Brown, Green, Longmans, & Roberts, London, 1858. His letter to Mrs Fry is a remarkably stoical and resigned account of the sorrow and grief he felt at the loss of three babies in a very short time. (baby Isabella and the twin boys.) 'I am thankful to say she kissed the rod on this as on former occasions with the sweetest Christian submission. "Them that sleep in Jesus shall God bring with him."' p.256
3. A. Parry, *Parry of the Arctic. The Life story of Admiral Sir William Edward Parry 1790–1855*. Chatto and Windus, London, 1963, pp. 212 and 262.
4. E. Parry, Rev., *Memoirs of Sir Edward Parry Kt. FRS* Longman, Brown, Green, Longmans & Roberts, London, 1853 p.263. The volume was printed for private circulation only but afterwards published by Hatchard and Sons Piccadilly. Proceeds of the 1st Edition were donated to the Royal Naval Female School at Richmond.
5. England and Wales, Quaker Birth, Marriage and Death Registers, 1578–1837.
6. A. Parry, *Parry of the Arctic. The Life Story of Admiral Sir William Parry. 1790–1855* Chatto and Windus, London, 1963, p.195. Lady Maria asked him to change the hours of departure for the Holyhead Packets to suit the 'passengers'.
7. Edward John Stanley was the brother of Sir John and twin son of Lord Stanley and a career politician. He entered Parliament in 1831 and became Undersecretary for the Colonies between 1833 and 1834. Head of the Home Office in 1834. Undersecretary for Foreign Affairs in 1846–1852. He was an unpleasant, caustic man nicknamed 'Benjamin Backbite' by some of his Whig party contemporaries. He died in 1869. D N B.
8. H.D. Traill. *The Life of Sir John Franklin RN* John Murray, London, 1896. 'Ignorance animated by self interest and disguised by typography.' p.145.
9. F. Fleming, *Barrow's Boys*. Granta, London, 1998, p.363.
10. A. Brandt, *The Man Who Ate His Boots. Sir John Franklin and the tragic history of the North West Passage* Jonathan Cape, London, 2011, pp.293–95. See also H.D. Traill. *The life of Sir John Franklin RN* John Murray, London, 1896.
11. G. Williams. *Arctic Labyrinth. The Quest for the North West Passage*. Allen Lane, London, 2009, pp.268–9.
12. TNA; ADM.7/187.no 4.
13. B. Lehane. *The North West Passage* Time Life Books, Amsterdam, 1981. An exploded diagram of the engines and the propeller is shown on pages 136–137.
14. C.E. Parry. *A record of her life; told mainly in letters*. Fletcher and Sons Norwich. Letter to Mrs Johnston May 10 1845, p.136.

Chapter 18

1. A. Brandt, *The Man Who Ate His Boots. Sir John Franklin and the Tragic History of the North West Passage* Jonathan Cape, London, 2011, and Professor G. Williams, *Arctic Labyrinth. The Quest for the North West Passage* Penguin Books, London, 2010. The latter gives a comprehensive list of sources and biographical studies in Part IV pp.404–6.
2. A. Brandt, *The Man who ate his Boots. Sir John Franklin and the Tragic History of the North West Passage* Jonathan Cape, London, 2001. This is an extremely well researched, even-handed account of his life the people who influenced his career, Lady Franklin and his sad death.
3. A. Parry, *Parry of the Arctic The life story of Admiral Sir Edward Parry 1790–1855* Chatto and Windus, London, 1963, p 220.
4. A. Brandt, *The Man who ate his Boots. Sir John Franklin and the Tragic History of the North West Passage* Jonathan Cape, London, 2011, p.304. The suspicion remains that much of the canned food turned out to be inedible because of a fault in its sealing.
5. G. Hutchinson, *Sir John Franklin's Erebus and Terror Expedition. Lost and found.* National Maritime Museum, Bloomsbury, London, 2017. This illustrated account produced for the NMM exhibition: *Death in the Ice; The shocking story of Franklin's Expedition*, is the most recent and fullest examination of the tragedy researched in collaboration with the Canadian Museum of History.
6. E. Belcher (Captain) CB RN. *The last of the Arctic Voyages. Being a Narrative of the expedition in HMS Assistance in search of Sir John Franklin during the years 1852–1853–1854.* 2 Vols. Discovery Books RGS.

Chapter 19

1. C.E. Parry, *A Record of her Life. Told chiefly in Letters* Letter dated 16 February 1845. To her children. Fletcher and Son. Norwich p.135. 'I (Parry) had successful surgical treatment for a stomach complaint. I desire to ascribe it all to His unmerited goodness, and to devote myself more entirely to His service who has dealt thus tenderly with me.'
2. E. Parry., Rev., *Memoirs of Rear Admiral Sir W. Edward Parry Kt FRS 1790 1855*. Longman, Brown Green, Longmans and Roberts, London, 1858, p.273. This appears to be the onset of a type of cancer.
3. Parry Family Archive. Letters to Lady Maria Stanley Alderley Park. 1847–1851.
4. J.B. Richardson, Dr FRS FRGS *A visit to Haslar* Journal of the Royal Naval Medical Services. Vol. 2 p.329
5. E. Parry, Rev., *Memoirs of Rear Admiral Sir W.E. Parry. Kt FRS* Longman, Brown, Green Longmans and Roberts, London, 1858, p.297. 'to his own family the five years of their life at Haslar present one unvarying picture of domestic enjoyment. For the first time since his residence at Port Stephens (Tahlee) he was enabled to combine official duties with the daily interests of the home circle.'
6. C.E. Parry, *A Record of her Life. Told chiefly in letters* Fletcher and Son. Norwich. Letter to Lady Parry. 31 January 1851, p.169
7. E. Parry, Rev., *Memoirs of Rear Admiral Sir W.E. Parry. Kt FRS* Longman, Brown, Green, Longman and Roberts. Appendices p.290.
8. C.E. Parry, *A Record of her Life. Told chiefly in letters* Fletcher and Son. Norwich. Letter to Edward Hankinson September, 1852, p.184
9. F. Fleming. *Barrow's Boys* Granta London. 1998 p.406 – 407
10. E Parry. Rev. *Memoirs of Rear Admiral Sir W. Edward Parry. Kt FRS 1790–1855* Longman, Brown, Green, Longmans and Roberts, pp.325–6.

11. E. Parry, Rev., *Memoirs of Rear Admiral Sir W. Edward Parry. Kt. FRS 1790–1855* Longman, Brown, Green, Longmans and Roberts p.328–9
12. *Times of London.* 21 October 1854.
13. C.E. Parry. *A Record of her Life. Told chiefly by letters* Fletcher and Son, Norwich. Letter to Reverend Hankinson on his birthday, 20 October 1853, p.191
14. This was part of the effort made by the British Government to suppress slavery which despite abolition continued down the West African coastline.
15. C.E. Parry, *A Record of her Life. Told chiefly in letters* Fletcher and Son, Norwich. Letter to Reverend R. Hankinson. Koblenz. 30 May 1855, p.210
16. C.E. Parry. *A Record of her Life. Told chiefly in letters* Fletcher and Son, Norwich, Letter from Rev. R.E. Hankinson to Reverend R. Hankinson. Greenwich, 20 July 1855, p.216

Chapter 20

1. G. Williams, *Arctic Labyrinth. The Quest for the North West Passage* Penguin, London, 2009. The cairn was built by Simpson and Dease on their boat voyage along the southern coast of King William Island p.341 This was the same route taken by Roald Amundsen in 1903–6 in *Gjoa* when he completed the entire passage.
2. The verse reads: 'Not here! The white North has thy bones and thou, heroic sailor soul, are passing on thine happier voyage now, Towards no earthly pole.'
3. R. Blake. *Religion in the British Navy. 1815–179. Piety and Professionalism* The Boydell Press, Woodbridge, 2014. 'The dissemination of truth and knowledge had become reality: the quest was quite often perceived in spiritual terms and subconsciously no doubt some officers of highly developed conscience might have welcomed a partial metamorphosis of the RN from an instrument of warfare to a protector of truth. This made it a profession to people of strong religious belief.' p.210
4. Ibid p.202
5. The rates of pay in the Arctic was double that for other commissions.
6. E. Parry, *Memoirs of Sir William Edward Parry. Kt FRS*, Longman, Brown, Green, Longman's and Roberts, London, 1858, p.296.
7. Sir John Barrow Bart. *An Autobiographical Memoir of Sir John Barrow Bart. Late of the Admiralty* John Murray, London, 1847, p.487. 'In fact these voyages have held out such fine examples of strict discipline without corporal punishment of kind treatment and wholesome indulgence on the part of the officers and in consequence therof of cheerful obedience exertion and alacrity on that of the men.'
8. Sir J. Barrow. Bart, FRS *An Autobiographical Memoir of Sir John Barrow Bart. Late of the Admiralty* John Murray, London, 1847.
9. E. Parry, *Memoirs of Sir William Edward Parry. Kt FRS*, Longman, Brown, Green, Longman's and Roberts, John Murray, London, 1858 p.280. Letter to Parry from John Franklin, 10 July 1845. 'I have left the magnetic observations of the *Erebus* to Crozier who is most assiduous respecting them. My share of the work at present seems to be more the training and overlooking of these gentlemen (his officers) than doing the work myself.'
10. He was especially anxious about hygiene. The regular airing of bedding was a daily instruction not easily carried out in confined spaces.
11. J.W. Croker, He had sent several letters complaining that Parry was too eager to comply with Barrow's Arctic schemes and did not possess sufficient independence of mind. Given that he was a serving officer and been selected by Barrow for the first expedition

with John Ross this is not surprising. Parry did defer to Barrow always and sometimes wrongly but he had little alternative.
12. S. Glaser. *The Spirit of Enquiry. Caleb Hillier Parry. MD FRS* Alan Sutton Publishing, Stroud, 1995. The papers included reports on Hay Fever, the first modern description, Facial Hemiatrophy, Thyyrotoxicosis, and Introductory Essays to the collections from unpublished medical writings of the late Caleb Hillier Parry 1825..
13. E. Parry. Rev., *Memoirs of Sir William Edward Parry Kt FRS* Longman, Brown, Green, Longmans and Roberts, London, 1858. His son Edward wrote that, 'in this second period of his married life he felt each day an increasing cause of thankfulness to Him who had thus permitted him to fill the void which had been left in his heart and home'. P.265
14. Sir John Barrow Bart. FRS *An autobiographical Memoir of Sir John Barrow Bart. Late of the Admiralty* John Murray, London, 1847, p.476.
15. T.H. Levene. *Science and the Canadian Arctic, 1818–76, from Sir John Ross to Sir George Strong Nares.* Arctic Magazine. Vol. 41.No 2 3/02/1988 pp.127–37.
16. Angel.M.V. see www.esdim.ncaa.gov/ocean. An holistic understanding of how the oceans work and a component of global studies.
17. Sir Clements Markham. KCB FRS, *Lands of Silence* Cambridge University Press, London, 1921.
18. Sir John Barrow Bart. FRS *Voyages of Discovery and research within the Arctic Region from the year 1818 to the present time by several Naval Officers employed by Sea and Land in search of the North West Passage from the Atlantic to the Pacific. With two attempts to reach the North Pole* John Murray, London, p.268.

Appendix 1
1. See A.R.T. Jonkers, *Earths Magnetism in the Age of Sail* John Hopkins University Press, London, 2003.frogpupp

Bibliography

Austen, J. *Emma; A Novel in three volumes* John Murray, London, 1816.
Austen, J. *Persuasion* John Murray, London,
Armitage, A. *The Life Story of William Herschel and his sister Caroline Herschel*, London, 1962.
Barrie, I. *Sextant: A Voyage guided by the Stars and the Men who Mapped the World's Oceans.* William Collins, London, 2015.
Barrow, Sir J. *An Autobiographical Memoir of Sir John Barrow Bart, late of the Admiralty, including reflections, observations and reminiscences at home and abroad from early life to advanced age* John Murray, London, 1847.
Barrow Sir J. *Voyages of Discovery and Research within the Arctic Regions from the Year 1818 to the present time under the command of several Naval officers employed by sea in search of the North West Passage from the Atlantic to the Pacific, with two attempts to reach the North Pole* John Murray, London, 1846
Beardsley, M. *Deadly Winter. The Life of Sir John Franklin.* Chatham. Rochester. 2002.
Blake, R. *Religion in the British Navy. 1815–1879. Piety and Professionalism.* Boydell Press, Woodbridge. 2014.
Blaug, M. *The Myth of the Old Poor Law and the Making of the New.* Journal of Economic History 23 1963.
Brandt, A. *The Man Who Ate His Boots; Sir John Franklin and the Tragic History of the North West Passage.* Jonathan Cape, London, 2011.
Brown, Dr R. *Chloris Melvilliana.* Royal Scottish Botanical Garden. Edinburgh.
Belcher, Sir E. *The Last of the Arctic Voyages Being a Narrative of the Expedition of HMS Assistance. Volume II.* Rediscovery Books, by kind permission of the R.G.S. Uckfield. UK. 2006
Berton, P. *The Arctic Grail. The Quest for the North West Passage and the North Pole, 1818–1909.* Viking Penguin. New York. 1988.
Biggs, A.S.A. *The Age of Improvement 1783–1867.* Longmans, London, 1979
Briggs. J. H. *Naval Administration 1827–1892,* London, 1897. *Across the top of the World. The Quest for the North West Passage.* British Museum Press, 1999.
Cavell, S.A. *Midshipmen and Quarterdeck Boys in the British Navy 1771–1831.* The Boydell Press, Woodbridge. 2012.
Cecil, D. *A Portrait of Jane Austen.* Penguin Books. Harmondsworth U.K 1980.
Conrad, J. *Geography and Explorers Last Essays.* London, 1926.
Cruickshank, D. & A. Burton. N. *Life in the Georgian City.* Penguin/Viking, London, 1990.
Cyriax, R.J. *Sir John Franklin's last Expedition.* Reprint. edition. Plaistow and Sutton Coldfield. 1997.
Dear, I. C. B. & Kemp. P. *Oxford Companion to Ships and the Sea.* (2ndEdit) OUP Oxford. 2005.
Dunn, R. *Navigational Instruments.* Shire Publications. Oxford. 2016.
Fisher. A. *Journal of a Voyage of Discovery to the Arctic Regions Performed Between 4th April and 18th November 1818 in H.M.S. Alexander.* 1819. London. Richard Phillips.

Fisher. A. *A Journal of a Voyage of Discovery to the Arctic Regions in HM ships Hecla and Griper in the years 1819 and 1820.* 1821. London. Longman, Hurst, Rees, Orme and Brown.

Glaser, S. *The Spirit of Enquiry. Caleb Hillier Parry.* MD, FRS Alan Sutton Publishing. Stroud UK 1995.

Good, G.A. *Sabine, Sir Edward.(1788–1883).*Oxford Dictionary of National Biography.10/012/2004.

Jackson, C.A.(ed.) *The Arctic Whaling Journals of William Scoresby the Younger. The Voyages of 1817, 818,and 1820.* Ashgate for the Hakluyt Society. Series III Vol. 21. 2009.

Holmes, R. *Coleridge. Early Visions.* Hodder &Stoughton, London, 989.

Holmes, R. *The Age of Wonder; How the Romantic Generation Discovered the beauty and Terror of Science.* Harper Press, London,2009.

Hurd, J.H. *A Chart of Baffin Bay with Davis and Barrow Straits by Captain J. Ross and Lieutenant W.E. Parry RN 1818–1819–1820 and the Discoveries of Captain Parry in 1822 and 1823 and Captain Lyon in 1824.*Admiralty Hydrographers Office. 1822.Engraved by J C Walker.

Hutchinson, G. *Sir John Franklin's Erebus and Terror Expedition. Lost and Found.* Adlard Coles Nautical. Bloomsbury. London2017.

Jardine, L. *Ingenious Pursuits.Building the Scientific Revolution.* Abacus Books. St Ives. 1999.

Jonkers, A.R.T. *Earths Magnetism in the Age of Sail.* John Hopkins University Pres, London, 2003.

King, R. Dr *The Franklin Expedition from First to Last* John Murray, London, 1855.

Lambert, A. *Franklin: Tragic Hero of Polar Navigation.* Faber &Faber, London,2009.

Lehane, B. *The North West Passage.* Time-Life Books. Amsterdam.1981.

Levere. T.H. *Science and the Canadian Arctic. A Century of Exploration.1819–1918.*Cambridge. 1993.

Loomis, C.C. *The Arctic Sublime.* In *Nature and the Victorian Imagination.* University of California. Berkeley. 1977.

Loomis, C.C. *The Arctic of the Imagination.* University of California. Berkeley.1978.

McGoogan, K. *Fatal Passage. The Untold story of Scotsman John Rae the Arctic adventurer who discovered the fate of Franklin.* Bantam Press, 2002.

Morriss, R. (ed). *The Channel Fleet and the Blockade of Brest 1793–1802.* Ashgate Press, Navy Records Society. Aldershot. 2001.

Markham, Sir C. R. *Lands of Silence.* Cambridge University Press London.1921

O'Reilly, B. *Greenland and the adjacent seas and North West Passage to the Pacific Ocean.* Baldwin, Craddock and Joy, London, 818.

Parry, A. *The Life Story of Admiral Sr Edward Parry 1790–1855.* Chatto and Windus, London, 1963.

Parry, E. *Memoirs of Rear Admiral Sir W. Edward Parry. Kt. FRS1790–1855.* Longmans, Brown, Green, Longman and Roberts, London, 1857.

Parry, W.E. *Parental Character of God.* Hatchard and Sons, London, 861.

Parry, A. *Parry of the Arctic: the life story of Sir Edward William Parry.1790–1855.* Chatto and Windus, London, 1963.

Parry, Sir W.E. Capt. RN FRS *Three Voyages for the Discovery of a North West Passage from the Atlantic to the Pacific and the narrative of an attempt to reach the South Pole.* Two Volumes. Harper & Brothers New York. 1840 New York.

Parry, Sir W.E. Capt. RN FRS *Thoughts on the Parental Character of God.* Privately Published.1841.

Parry, Sir W.E. Rear Admiral. RN FRS *A Lecture on the Character, Condition and Responsibilities of British Seamen. Given at the Victoria Rooms, Southampton. 19/12/1853.* Harrison and Son, London, 1854.

Richardson, J. Dr MD, FRS, *Fauna Borealis; Natural History of the Arctic Regions,* London,

Poulsom, N.W. & Myres. J.A.L. *British Polar exploration and research: a historical and Medallic record with biographies.1818- 1999.* Savannah, London, 2000.

Rodger, N.A.M. *The Command of the Ocean. A Naval History of Britain, 1649–1815,* Allen Lane, London, 2004.

Ross, M.J. *Polar Pioneers. Sir John Ross and James Clark Ross,* Montreal, 1994.

Sabine, Sir Edward. FRS, *An account of experiments to determine the figure of the earth by Means of the pendulum vibrating seconds in different latitudes; as well as on various other subjects of philosophical enquiry.* John Murray, London, 1825.

Sabine, Sir Edward. FRS, *Observations on days of unusual Magnetic disturbance Made at the British Colonial Observatories under the Departments of Ordnance and Admiralty.* Longman, London, 1843.

Savours, A. *The Search for the North West Passage.* Chatham. Rochester.1999.

Scoresby, W. Jun. *An Account of the Arctic Regions: with a History and Description of the Northern Whale Fishery. 2 Vols.* Archibald Constable and Co, Edinburgh, 1820.

Skelton, R. A. *Explorers Maps. II. The North West Passage. Frobisher to Parry.* Geographical Magazine Vol.26 No.4.

Smith, E.C. OBE. RN Capt. *A Short History of Naval and Marine Engineering.* Cambridge University Press, London, 1938.

Smith, E.C. OBE RN Capt. *The Centenary of Naval Engineering. A review of the Early History of our Steam Navy.* Paper published by Newcomen Society for the Study of the History of Engineering and Technology. 1921.

Spufford, F. *I May be Some Time.* Faber and Faber, London, 1996.

Shelley, M. *Frankenstein.* or *The Modern Prometheus.* Revised Edition. Penguin Books, London,

Simpson, T. *Narrative of the Discoveries on the North Coast of America effected by the officers of the Hudson's Bay Company during the year 1836–1839.*

Traill, H.D. *The Life of Sir John Franklin* John Murray, London, 1896.

Williams, G. *The Quest for the North West Passage.* Penguin, London, 2009.

Williams, G. Quilley G. Arutiunov, Forgan S. *Smoking Coasts and ice-bound Seas: Cooks Voyage to the Arctic.* Catalogue to the exhibition at the Captain Cook Memorial Museum. Whitby.2008.

Magazines and Periodicals
Glasgow Morning Herald 23 July 1826.
The Times London 21 October 1854.
Polar Record (5) 1948 pp.348 – 350.
The Sun (London) 19 June 1855.
Berkshire Chronicle 5 June 1847.
Blackwood's Edinburgh Magazine Vol. 81 (497) March 1857.
History Today Vol.10. No 2. P.100. (1960) W. Gillies Ross. Parry's Second Voyage.

Primary Sources
NATIONAL MARITIME MUSEUM N.M.M. Caird Library, Greenwich, England.
AG C/10/16. Admiral Sir William Edward Parry: Holograph to Sir John Franklin 28 February 1845.

N.M.M. 910.4.(987) 1818; 094. A Voyage of Discovery made under the order of the Admiralty in His Majesty's Ships *Isabella* and *Alexander* for the purpose of exploring Baffin Bay and enquiring into the probability of a North West Passage. John Ross K.S. Longman, Hurst, Rees, Orme, and Brown, London, 1819.
AG C *C*21 MS 78038. T.C Croker, Letter to *London Gazette*. Outspoken denunciation of Captain W.E. Parry
PUBLIC RECORDS OFFICE. (PRO) Kew. England.
AG C /10/*C*/21. Letter from Thomas Crofter Croker discussing Naval Officers on Surveys. 1828.
AD M 101/103/5. Medical and Surgical Journal byAssistant Surgeon of HMS *Griper*. 1824.
AD M 55/5. HMS *Alexander*. Rates of Chronometer days work.
Supplementary Logs and Journals.1818–1819.
SCOTT POLAR RESEARCH INSTITUTE. SPRI. Cambridge England.

Sir William Edward Parry Collection.
GB /015/GB15. Sir William Edward Parry (1790–1855.)
MS 438 /26/247 – 322. Letters to Mother. Sarah Parry.
MS 248 45210-D. Letter to Franklin 1823.
MS 438/26/34. Letters to Sir John Franklin.
MS 578/1 Letter to Mother. Sarah Parry.
MS 438/323 – 336. Letters to Sisters.
MS 430/*26/6* – 63 D. Letters to Caleb and Sarah Parry.
MS 438/26/6. Letter to Parents.
MS 438/26/5. Letters to Father. Caleb Parry.
MS 438/26/148 – 209. Letters to Brother. Charles Parry.
MS 438/26/34. Letter to John Franklin.
MS 248/452/10 – D. Letters to John Franklin.
MS. 438/41/1 – 21D. Miscellaneous Papers.
MS 438/5. Watercolour Sketches. Coastal Recognition. 1810–1815.
MS 1199/1/1. Private Journal of W.E. Parry. HMS *Alexander*, 1818.
MS 1059. Private Journal of Captain J. Ross (Ross Expedition 1818). 3 Volumes.
GB /015/GB15. Sir William Edward Parry (1790–1855)

Isabella Louisa Parry Collection (1813–1839)
GB 015. Family Papers.
MS 438/25/24/-27. Letters to Half Sister.
MS 438/294/-296. 4 Letters to sisters in law concerning general news from Australia. Dated15/12/1832.to28/5/1833.MS438/25/28. Letter to Sir John Stanley.
MS 438/25/294 – 296. Letter to Caroline Martineau, sister in law.
MS 438/25/297. Letter to husband. Undated.
MS 438/25/13–19. Letters to mother-in-law. Lady Maria Stanley. Various dates.
MS 438/25/293. Letters to sister-in-law. Gertrude Parry. Undated.

Sir John Barrow Collection (1818–1843)
GB 015.
Sir John Barrow. First Baronet Secretary of the Admiralty.
General Correspondence.

Sir Edward Sabine. (1819–1858)
Special Collection 015/ 528.2. The North Georgia Gazette and Winter Chronicle.
GB 15. MS 549; SL.Sir Edward Sabine. Meteorological Journal.11th May to 6 September 1819. 1 Journal.
MS 1226/27/1 – 3D. Letters to Sir James Clark Ross 1858 regarding paper on magnetic observations at Port Leopold.
MS 1243/1-4; M J. Correspondence held at PRO
MS 549; SL Meteorological Journal, 11 May – 6 September 1819. (HMS *Hecla*.) 1 Volume.
MS248/316 Franklin letter to Sabine.

SPRI Picture Library Database.
Drawings from Parry's Arctic Expedition of 1821–23 printed by John Murray Negatives. Art. Collection 1550,1543,1546,1562,1545,1551,1558,1555.

The Royal Society, London, England.
Proceedings of the Royal Society. 30 November 1855.
Philosophical Transactions. No 16. 1826.
Quarterly Review. J. Barrow. FRS *On the Polar Ice and a Northern Passage into the Pacific*, pp.199–223.
Blackwood's Edinburgh Magazine Arctic Adventure. Vol. 81. (497) March 1897.
Blackwoods Edinburgh Magazine November 1820, pp 219–23
Royal Geographical Society, London, England
RGS. SC/73/1. Lieutenant E. Hooper. Private Journal of the Voyage of the *Isabella* and *Alexander* in search of a North West Passage. March to November 1818. Ross Expedition. MS Quarto. 225p.
RGS. SC/73/2. Lieutenant E. Hooper. Private Journal of the Voyage of HMS *Hecla* and HMS *Griper* under the command of Commander W.E. Parry. 1819–1820.r gsu 213332.

Yale Centre for British Art. Bedford Square, London, England.
Folio B2007/10. 1821–1822 G.F Lyon. 40 Evocative Sketches of Inuit Life. Melville Peninsula and Hudson's Bay. Graphite and Yellow Wash.
G 650 1821–25. Captain G.F. Lyon.1795–1832. Author, illustrator. The Private Journal of Captain George Francis Lyon of HMS *Hecla* during the recent voyage of Discovery under Captain Parry. 468 pages. 7 leaves of plates.

Australian National University. Canberra. Australia.
Australian Dictionary of Biography. Volume 2 (MUP) 1967

Mitchell Library. University of New South Wales. Sydney. Australia.
Call No. SAFE/PXE 1072.Wallace. J. Album of original drawings Captain James Wallis and Joseph Lycett ca. 1817–1818, bound with Agricultural Historical Account of the Colony of New South Wales, London, Rudolph Ackermann 1821.Reference code 954703.
The Royal Australian Historical Society Vol.IX.Part III.1923. J.S. Campbell. First decade of the Australian Agricultural Company. 1824–1834.
Journal and Proceedings (Royal Australian Historical Society) Vol 9 Part 3 1923 pp.113–60
Kelvin Smith Library. Department of Special Collections. Cleveland Ohio. Copy of Original Frontispiece for Pamphlet. 'Voyage of Discovery by the renowned Captain Munchausen GCBA'. 'Munchausen at the Pole'

Columbia University New York.
Butler Library.
University of Southampton. UK.
MS. 45 A 01876/1 Annotated copy of GF Lyons' Private Journal. 1824.
Online Sources.
www.historyhome.co.uk/people.htm
Glasgow Morning Herald. 27th July 1826.
The Mariner's Mirror. Volume 104. 4 November 2018 www.web.archive.org/web 20071214085254/
The Letters of Sir Joseph Banks. www.scientific.co. 89.168.217.31
www.cottontimes.co.uk/poorlawo.htm. 'Workhouse' A fact of life in the Industrial Revolution 14 December 2007. University Of Toronto.
Parry, Sir William Edward. *Dictionary of Canadian Biography*. Vol. VIII. 1851–1860. www. biograhi.ca/en/bio/parry_William_Edward_8 E. html
c.davies@kesbath.com. Archive of King Edwards Grammar School, Bath. http://historicengland.org.uk/listing/selection-criteria/listing-selection/ihas-buildings/
Native Land Digital/Native – Land. CaArctic Indigenous Peoples in Canada. Milton M.F. Freeman. Updated by Z. Parrott, M. Felice. Published online 15 March 2007. Last edited. 24 October 24 2017.
PRO Kew. Currency Converter 1830 -2017.

Picture Sources.
Parry family Archive.
Fig. 1.Mezzotint of Sarah and Caleb Parry. Date unknown.
Fig. 2. Line and wash Drawing of Summerhill c.1860.
Fig. 13. Isabella Stanley with their two eldest sons Edward and Charles. Date unknown. Engraving. Fig. 18. Catherine Hoare with her two daughters, Louisa and Priscilla 1839. Watercolour.
Fig. 19 Sir W.E. Parry FRS 1842. G. Richmond. Pen and ink drawing.
Front Cover. *Commander W.E. Parry RNG Drummond 1821* Oil on Canvas. Family collection.
Royal Geographical Society.
Fig. 14. The wreck of the *Fury*. H. Head RN Engraving by E. Finden.
Map 1. New Map of the United States of America with the British Dominions on that Continent. Cf 1790 Samuel Dunn.
Map 2. The Northwest Passage showing the Coast prior to 1819 and the Discoveries Of Sir Edward Parry and subsequent Discoveries. Coloured Engraving. E. Welley.
Royal Geographical Society Library.
Map 1 (b).Gerardus.Mercator. Polus Arcticus. C 1641.
Fig. 3. *Wreck of the Magnificent on the Boufoloc Rock. Oil on Canvas* J. Schekty. 1804.

Scott Polar Research Institute.
Fig. 8. *Sir* John *Franklin* Oil on canvas. E. Lewis c.1832.
Fig. 10. *His Majesty's ships* Hecla *and* Griper *blocked up by ice in winter harbour* F. Beechey RN Watercolour drawing 1820.

Mitchell Library, Sydney, New South Wales.
Fig. 16 *Port Jackson c.1826*. J. Wallis, J. Lycett. Colour engraving.
Fig. 17. *Newcastle NSW* 1828 Artist unknown. Black and white engraving.

Royal Society of London.
Fig. 6. *Sir Joseph Banks KB* William Daniell cf. 1803. Oil on canvas.

The Royal Australian Historical Society, Vol.IX.1923.
Map 4. Port Stephens and the original Grant of Land of the AAC.
Map 5. Parry's expedition through the Liverpool range across the Liverpool Plains and the Pel River March–April 1832.

National Portrait Gallery, London,
Fig. 15. *Portrait of Jane, Lady Franklin. Age* 24. A. Romilly.
Fig. 4. *Portrait of Sir John Barrow* Oil on Canvas. G.T. Payne.
Fig. 7. *Sir John Ross* Oil on canvas. J Green. cf 1829.

Mellon Institute for British Art.
Fig. 13. *Canoe of the Savage Islands 1823* Captain G.F. Lyon RN. Coloured sketch.

Collection of R. Edinger. USA.
Fig. 9. *Ross and Parry introduce themselves to the Inuit at Elah. Western Greenland.* Captain J. Ross RN Coloured sketch. 1819.

Index

Alderley Park, Cheshire, 97
Alexander (brigantine), 32–6
America, United States of, 17
Amiens, Treaty of, 11
Amphrite, HMS, 171–2
Amundsen, Raold, 169
Archimedes, SS, 155
Arctic expedition, Ross's, 32–50
Aurora Borealis (northern lights), 25, 56–7
Austen, Frank, 14
Austen, Jane, 5–6
Australian Agricultural Company, 119, 123

Back, Lieutenant George, 83
Baffin Bay expedition, 31, 35–43
Baffin, William, 35
Baker, Captain, 15–16
Banks, Sir Joseph, 6, 23–5, 27
Barents Sea, 24
Barker, Henry, 31
Barrow, John, 25–8, 44, 48
Barrow Strait, 51
Barton, Mr (NSW accountant), 129, 131, 134, 139–40
Bathurst Bay, 40
Beaufort, Captain Francis, 120, 122
Beechey, Lieutenant Frederick, 41, 46, 56, 87
Beechey, Sir William, 47, 65
Bermuda, 18–19
Blockade against France, sea, 9–16
Blossom, HMS, 87
Blue light ships, 11
Boiling station, 74
Bounty, Cape, 52
Bounty, Parliamentary, 36, 52, 64, 100, 107
Bourke, General Richard, 134
Bowen, Port, 88
Brisbane, Governor Thomas MacDougall, 123

British and Foreign Sailors Society, 172
British Guiana, 7–8
Browne, Miss, 48–9, 64
Brunel, Isambard Kingdom, 153–6
Buchan, Captain David, 28, 32, 41
Burford, Robert, 31
Burnett, William, 122
Burney, Fanny, 6
Bushman, Charles, 69
Bylot peninsula, 61
Byron, Lord, 63

Caledonian Canal, 160–1
Canal cutting, 53
Carlo (dog), 57
Carrington, New South Wales, 125–9
Cary Island, 37
Chenery, Miss (governess), 159–60
Clarence, Duke of, 109–11, 117–18
Clothing, expedition, 32
Cockburn, Sir George, 44–6
Coleridge, Samuel Taylor, 6
'Continental System', 17
Convicts in Australia, 125, 128–31, 137–9
Cook, Captain James, 23–4, 85
Coote, Commander Robert, 171
Coppermine River expedition (Franklin), 41, 47, 53
Corn Laws, 142
Cornwallis, Admiral Lord William, 9–11, 14–15
Court martials, 93
Cowper, William, 145
Cresswell, Lieutenant Gurney, 174
Croker, John Thomas, 29, 115–18, 153
Croker, John Wilson, 29, 51–2, 95
'Croker Mountains', 39, 42–3, 50
Crozier, Captain Francis, 91, 100, 104, 164–5
Cunard Line, 151

Dangar, Henry, 130, 135–6
Darch, Henry, 122
Darling, Sir Ralph, 124, 133–4
Davy, Sir Humphry, 5, 83
Dawson, Robert, 123
Drummond, Samuel, 65
Dumaresq, Colonel Henry, 139–40
Dundas, Henry, 9, 25
Dundas, Robert Saunders *see* Melville, Viscount

Eardley-Wilmot, John, 34, 37, 45, 66, 122
Ebsworth, James, 126–7, 131–2, 134
Ebsworth, Thomas, 137
Edwards, John, 46, 79
Ellesmere Island, 38
Ems, Germany, 175–6
Enterprise, HMS, 167, 173
Enterprise ice boat, 104–105
Erebus, HMS, 164–5, 178
Ericsson, John, 155

Fido (dog), 103–104
Field, Joshua, 154, 156
Fisher, Alexander, 46, 64
Fisher, Reverend George, 69, 71–2, 87
Fitzjames, Captain James, 164, 168, 178
Flinders, Matthew, 23
Flogging, 19, 110, 180
Foodstuffs, expedition, 33
Foster, Henry, 87
Frankenstein (Mary Shelley), 63–4
Franklin, Jane, 26, 47, 85, 121, 168–9
Franklin, Sir John, 46–7, 73, 82–3, 91–2, 180–1
  escapes wounding, 15
  Spitzbergen expedition, 28, 33, 41
  Fury, HMS, 67–91
  offered Greek mission, 133
  in Tasmania, 161–2
  knighted, 120–1
  final expedition, 163–9, 178
  death, 166–8

Gardner, James Anthony, 12
Geographical Society, 26
Geological Society, 5
Geology, 5
Geomagnetism, 23, 30, 34

George IV, King, 64
Ghent, Peace of, 18
Gjoa yacht, 169
Goode, Reverend Ambrose, 144–5
Gorgon, HMS, 154
Göttingen University, 6–7
Graham, Sir James, 154, 166, 172
Grape cultivation, 139, 141
Great Western, SS, 153–4
Greenland, 37, 49
Griffin, Jane *see* Franklin, Jane
Griffiths, Admiral, 19–20
Griper, HMS, 45–6
  see also Northwest Passage, WEP's first voyage

Haddington, Lord, 163–5
Halifax, Nova Scotia, 17–18
Hall, Henry, 130, 137
Hall, Sir Charles, 130, 133–4, 137, 155
Halse (clerk), 74–5
Hammerfest, Norway, 104
Hankinson, Catherine, 94
Hankinson, Edward, 176
Hanley, Edward, 103
Hardcastle, Edward, 171
Hardy, Sir Thomas, 153
Head, Horatio Nelson, 87, 91
Hecla and Griper Bay, 52
Hecla, HMS, 44–62, 67–91
  see also Northwest Passage, WEP's voyages
Herschel, Sir Thomas, 28
Herschel, William, 5
Hoare, Catherine *see* Parry, Catherine
Hoare, Samuel, 160–1
Hobhouse, Sir Benjamin, 4, 6–8, 27
Hood, Midshipman, 82
Hooper, Lieutenant Edward, 40–1, 45–6
Hooper, William, 49, 87
Hope, Sir George, 28
Hoppner, Lieutenant Henry, 42, 56, 87, 93
Hunter Valley, New South Wales, 139, 141
Hurd, Captain Thomas, 116
Hydrographical research, 23
Hydrography, 23

Icebergs, 25
Ice masters, 30, 34

Index

Igloolik island, 77, 79
Illigliuk (Inuit woman), 75
Inuit people, 37, 72, 74–8, 80–2
Inventories, expedition, 32
Investigator, HMS, 59, 167, 173–4
Investigator ice boat, 104–105
Isabella (whaler), 32–3

Jenner, Edward, 4–7
Jeppings, Sir Robert, 103
Jervis, Admiral John, Lord St Vincent, 10–11

Kater's Pendulum, 33–4
King, Dr Richard, 165
Kotzebue, Lieutenant (Russian Navy), 24

Lambton cutter, 125–6
Lancaster Sound, 38–40, 42–53
Lands of Silence (Sir Clements Markham), 184
Lerwick, Shetland Islands, 80
Liddon, Lieutenant Matthew, 46, 61
Life aboard ship, 11–13
Life preserver, 48–9
Lights, navigation, 156
Lloyd, Thomas, 156
Longitude Act 1775, 36
Loutherbourg, Phillip de, 31
Lyon, Captain George, 67, 69, 71, 80–1

Macarthur, John, 123–5
Magnetic variation, 30, 36–8, 40, 89, 100, 147
   see also North Pole
Markham, Sir Clements, 65
Martineau, Joseph, 87
Martin, Sir Byam, 32
Martyr, Charles, 18
Maxwell, Sir Murray, 21, 27, 29, 44, 65–6, 139
McClintock, Leopold, 175
McClure, Captain Robert, 59, 173–4
Mechanisation of agriculture, 142
Melville Island, 51
Melville, Viscount, 27–30, 34, 42, 85
Mercator, Gerardus, 35
Merchant Seaman's Act, 1835, 172
Messages in bottles, 35–6

Middleton, Christopher, 71
Mirages, 39–40
Mitchell, Major (Surveyor General), 136
Montagu, John, 162
Muirhead, Captain, 37
Murray, John (publisher), 64, 83–4, 100, 111

Naming convention, 51
Nansen, Frederick, 109
Nares, George, 183
NASA, 36
Nautical Astronomy (William Parry), 17
Nautilus (ship), 71
Naval and Military Bible Society, 172
New South Wales, 123–41
North Pole, 99–109, 146–8
   see also magnetic variation
North Pole, position of, 23
Northwest Passage:
   search funded by Royal Society, 23
   WEP's first voyage of discovery, 44–62
   WEP's second voyage of discovery, 67–91

Observatory, portable, 54
Oil, whale, 27
Oomiaks (Inuit canoes), 72
Orkney Islands, 40, 71, 107–108
Over-wintering, 53–8

Packet postal services, 150–2
Parhelia, 58
Parker, Sir William, 156
Parry, Caroline (WEP's sister), 3, 6–7, 87–8
Parry, Catherine (WEP's wife), 160–1, 172–3, 176, 182
Parry Channel, 53
Parry, Charles (WEP's brother), 3, 6–8, 93–4
Parry, Charlie (WEP's son), 171–2, 175
Parry, Dr Caleb Hillier (WEP's father), 3–5, 19, 21, 67
Parry, Emma (WEP's sister), 34
Parry, Frederick (WEP's brother), 3, 6–8, 13–14
Parry, Isabella (WEP's wife), 83–4, 88, 95–9, 101–103, 157, 182

Parry, Sarah (WEP's mother), 3, 134
Parry, Sir John Franklin, 120
Parry, Sir William Edward:
  born, 3
  schooling, 3–4, 6
  joins Royal Navy, 11
  awarded doctorate from Oxford University, 121–2
  on:
    HMS *Alexandria*, 16–17
    HMS *La Hogue*, 18
    HMS *Maidstone*, 18
    HMS *Niger*, 20
    HMS *Tribune*, 15–16
    HMS *Vanguard*, 16
    HMS *Ville de Paris*, 11–14
  commands:
    Alexander, 35–44
    HMS *Carron*, 18
    HMS *Fury*, 67–91
    HMS *Hecla*, 44–62
  promotions:
    midshipman, 11
    lieutenant, 16
    first lieutenant, 18
    captain, 63
    Rear Admiral of the Blue, 173
  applies for expeditionary voyage, 23
  expedition to Baffin Bay, 35–43
  first voyage for the discovery of a North-West Passage, 44–62
  second voyage for the discovery of a North-West Passage, 67–86
  third voyage for the discovery of a North-West Passage, 86–91
  final Arctic expedition, 99–109
  appointed to Captain Ross's Arctic expedition, 28
  appointed assistant hydrographer, 85
  as hydrographer to Royal Navy, 95, 115–18
  as New South Wales Commissioner, 119–41
  as Poor Law Commissioner, 144–6
  as Comptroller of Steam Machinery, 153–6, 170
  as head of Steam Packet Service, 153–6, 170
  as Superintendent of Haslar Hospital, 170–3
  as Lieutenant Governor Greenwich Hospital, 175
  first experience of Arctic, 17
  marries Catherine Hoare, 160
  marries Isabella Stanley, 97–9
  daughter, Emmeline, born, 145–6, 152
  daughter, Isabella, dies, 145–6, 149
  daughter, Lucy, born, 134
  son, Stanley, born, 111
  twins born, 126–7
  children with Catherine born, 160
  and Franklin's last expedition, 167
  friends with ice masters, 30
  and geology, 18
  and George Lyon's expedition, 94
  journals, 64, 67, 77, 81, 84–6, 90
  knighted, 120–1
  moves to:
    3 Downing Street, London, 46
    49 Weymouth Street, London, 115
    Congham Lodge, Norfolk, 145
    Devonshire Place, London, 151–2
    Mattishall, Norfolk, 145
    Mount Edgecombe, Kent, 157
    Northbrook House, Hampshire, 173
    Tahlee, New South Wales, 129–30
  Nautical Astronomy (pamphlet), 17
  opposition to flogging, 110, 180
  portrait by Sir William Beechey, 47, 65
  praised by Cornwallis, 15
  meets Sir Joseph Banks, 27–8
  meets John Barrow, 28
  and John Franklin, 82–3
  preparation for voyages, 28
  presented with Freedom of Bath, 67
  elected Fellow of Zoological Society, 152
  raises educational standards, 58
  relationship with Ross, 30, 38
  returns from New South Wales, 141–2
  testimonial dinner by Australian Agricultural Company, 143
  concern for seamen's conditions, 19–21
  religious conviction, 49–50, 58, 86, 118, 138, 172, 180
  attention to detail, 181
  fondness of Lake District, 173
  romantic attachments, 19, 83–4
  salary, 119–20, 144
  and violin playing, 6, 16, 18

illness, 89–90
and Isabella's death, 157–8
death, 176–7
grave, 178
lack of epitaph, 179
Peninsula Steam Navigation Company, 151
Penrhos, Anglesey, 97, 118, 151, 157
Persian (ship), 141–2
Philosophical Society, Bath, 4–5
Plays on board, 54–5, 74–5, 78–9, 89, 180
Polar bears, 38
Polus Arcticus map (Mercator), 35
Poor Law, 142–6
Porden, Eleanor, 83, 85, 88
Portsmouth Sailors Home, 172
Port Stephens, New South Wales, 125–7
Priestley, Joseph, 4–5
Prince Regent Inlet, 51, 80, 88
Prismatic cross, 86

Rae, Dr John, 167, 175
Rawdon, Francis, 87
Reid, George, 69
Reindeer, 101, 104
Repulse Bay, 73
Resolute, HMS, 174
Richardson, Sir John, 73, 82–3, 167, 171
Richmond, George, 65
Ricketts, Captain, 9, 11, 13
Robertson, George, 19
Ross, Captain John, 26, 29–43, 45, 48, 91–2, 146–8
Ross, Lieutenant James Clark, 40–1, 46, 68, 87, 107
Royal National Lifeboat Institution, 172
Royal Navy, size of, 23
Royal Society of London, 6, 23
'Rule of the road' at sea, 156

Sabine, Colonel Richard, 30
Sabine, Sir Edward, 33–4, 38–41, 43, 55–6, 68
Sacheuse, John, 33, 37, 42
Savage Islands, 72
Scoresby, William, 24–5, 28, 101, 105, 108
Scott, William, 61

Scurvy, 10, 33, 55, 78
Settlement, Law of, 142
Sheep breeding, 4–5
Sherer, Joseph, 87
Sledging, 75–6
Smith, Francis Petit, 155
Snow blindness, 59
Soest, Dr, 176
Sophia Jane steamboat, 131
Southampton Island, 72–3
Spitzbergen, 28, 33, 41, 104
Stanley, Lord Edward, 161–2
Stanley, Reverend Edward, 96, 98
Stanley, Emmeline, 145–6, 158–9
Stanley, Isabella *see* Parry, Isabella
Stanley, Sir John, 95–6
Stanley, Lady Maria, 95–8, 145, 159–60
Steam power, 150–6
Stoves, Sylvester, 74, 90
Stroud, New South Wales, 129, 136–8
Summerhill, Bath, 5, 182
Symonds, Captain William, 153

Tambula, Mount, Java, 28
Tasmania, 134, 161–2
Terror, HMS, 164–5, 178
The Prince of Wales whaling ship, 167
Thoughts on the Parental Character of God (William Parry), 159, 182–3
Tinned food, 33
Trafalgar, Battle of, 14
Trek, Arctic, 104–108
Tribune, HMS, 15

Valorous, HMS, 154
Van Diemen's Land *see* Tasmania
Ville de Paris, HMS, 9, 14
Villeneuve, Admiral, 14

Warren, Sir John, 17
William (ship), 122
Wilmington, John, 26
Winter Chronicle and North Georgia Gazette (newspaper), 55
Winter Harbour (Melville Sound), 53–61
Winter Island, 74
Wordsworth, William and Dorothy, 6, 173

Dear Reader,

We hope you have enjoyed this book, but why not share your views on social media? You can also follow our pages to see more about our other products: facebook.com/penandswordbooks or follow us on X @penswordbooks

You can also view our products at www.pen-and-sword.co.uk (UK and ROW) or www.penandswordbooks.com (North America).

To keep up to date with our latest releases and online catalogues, please sign up to our newsletter at: www.pen-and-sword.co.uk/newsletter

If you would like a printed catalogue with our latest books, then please email: enquiries@pen-and-sword.co.uk or telephone: 01226 734555 (UK and ROW) or email: uspen-and-sword@casematepublishers.com or telephone: (610) 853-9131 (North America).

We respect your privacy and we will only use personal information to send you information about our products.

Thank you!